CARNIVOROUS PLANTS
of the United States and Canada

Sarracenia flava

JOHN F. BLAIR, *Publisher*

Winston-Salem, North Carolina

CARNIVOROUS PLANTS

of the United States and Canada

by Donald E. Schnell

Library of Congress Catalog
Card Number: 76–26883
ISBN 0–910244–90–1
Printed in the United States of America
by Lebanon Valley Offset Company Incorporated

Library of Congress Cataloging in Publication Data

Schnell, Donald E   1936–
    Carnivorous plants of the United States and
Canada.

    Bibliography: p.
    Includes index.
    1. Insectivorous plants—United States. 2. In-
sectivorous plants—Canada. I. Title.
QK917.S36   583'.121'0973   76–26883
ISBN 0–910244–90–1

Sarracenia psittacina

For Lee Anne
& Kristen

Foreword

By C. RITCHIE BELL
Professor, Department of Botany,
University of North Carolina, Chapel Hill
and Director, North Carolina Botanical
Garden, Chapel Hill

Since their discovery by botanists over two hundred years ago, the world's carnivorous plants have been the center, from time to time, of much scientific study and public interest. They have also been the center of a number of misconceptions and, in the past few years, of considerable commercial exploitation.

Carnivorous plants are both colorful and biologically interesting, but they are *not* house plants. Their specific environmental requirements often involve very high humidity, high light intensities, quite acid soil, pure water, and seasonal temperature changes; such requirements are *not* met by a pot of generalized dirt, treated tap water, and the usual temperature and humidity ranges of the average home, office, or apartment! It is likely that not one "cultivated" carnivorous plant in a thousand lives a full year and probably less than one in ten thousand survives for two years in the hands of most amateur "collectors" or those who, unfortunately, are misled by the commercial advertisements for these unique plants and attempt to grow them, merely as a novelty, on the kitchen window sill.

Given the above background, this book becomes something of a landmark in carnivorous plant publication. Not only does it bring together in a very readable fashion the many interesting botanical facts concerning the form, function, and ecology of our carnivorous plants, but the author makes a strong, and very timely, case for the preservation of these unique members of the plant world from rapid extinction through a combination of realistic protection from continued collecting, conservation of their specialized habitats, responsible horticultural research, and the general cultivation of only those plants actually grown from seeds or, where possible, cuttings.

C. R. B.

July, 1976

Preface

This book is intended for practical use. It is not conceived of as a coffee table or bookcase ornament; the entire plan and structure of the work is centered around utility.

The photographs have been chosen with a view toward sharpness of important features, clarity, color fidelity, and most of all, how well they represent the plant. You will find only a few drawings in this book, and these are mainly of a figurative nature. The trouble with using only botanical drawings for identification is that they represent a two-dimensional, colorless average of characteristics in an ideal plant that rarely exists. This presents a problem for the beginner in the field, who has a book of drawings in hand but sees no plants that resemble any of the drawings, or perhaps sees too many.

The photos are intended to show the plants as they really are, and most of the pictures have been made in the field. If some of the photographs seem pretty or artistic, that is just a bonus. Since no view of a plant or group of plants can show all the important characteristics one might wish to see, I have presented multiple views of them where necessary.

Occasionally, in the field, where the natural backgrounds tended to camouflage, I have had to resort to the use of a neutral gray background, reflectors, or ancillary lighting to make the subjects stand out properly. Only a few pictures were made under studio conditions, using live plants from my collection. After having observed, grown, and worked with carnivorous plants for eighteen years, I felt capable of making a proper selection of cultivated plants for photography.

I have departed from the traditional botanical scheme of presenting species within a genus in alphabetical or other order, preferring to start with some common, more readily observed plants or those that would best illustrate general characteristics, or perhaps grouping similar species together so that comparisons could be seen.

The carnivorous fungi have been excluded. The book deals only with the green seedplants. Fungi require culture and microscopes for proper observation, and the species are in a state of taxonomic disarray at present. While fascinating and certainly deserving of further study, they are not ready for this sort of presentation.

The text is not complete in the classical botanical sense of a monograph, but it was not intended to be. On the other hand, the occasional or beginning naturalist may find more than he needs and can easily be selective in what he reads. Those interested in further study will find references on many levels at the end of the book.

. . .

I would like to thank John F. Blair, Publisher, and his staff for taking on this project in the first place, and then seeing it through to a quality production. Warren P. Stoutamire was kind enough to read portions of the text and offer many useful suggestions. Letters and discussions with Peter Taylor and Katsuhiko Kondo were very helpful to me in developing my tentative concepts of the difficult taxonomy of *Utricularia*, but I bear full responsibility for the system presented in Chapter 7, including errors. Finally, I offer tenderest appreciation to my wife Brenda, who has been patient and encouraging during the preparation of this book and who has accompanied me on many field trips, often serving as my stern early-warning system for potentially dangerous reptiles, quicksand, and treacherous bogholes.

Contents

I. Carnivorous Plants—An Introduction

The concept of a plant that traps and partially consumes small animals was suggested and studied long before 1875, when Darwin's book *Insectivorous Plants* appeared. In that volume Darwin correlated a great deal of the knowledge of his day and added the results of his own numerous experiments and observations. Since then, interest in carnivorous plants has grown remarkably. Concurrently with this growth, new discoveries and ideas have arisen, and these have further increased interest in carnivorous plants. Imaginations have also soared, but stories of giant or man-eating plants have proved to be entirely mythical.

Green plants can derive a large part of their chemical requirements for sustaining growth and reproduction from some very basic but essential elements. These include energy from sunlight, oxygen and carbon dioxide from the air, and water and certain minerals from the earth or water in which the plants grow. Through photosynthesis, in which green chlorophyl plays so large a part by transforming sunlight into chemical energy, carbohydrates are built up from water and carbon dioxide. These various carbohydrates themselves are used as energy sources and building blocks for synthesizing myriad other chemical materials needed by the plant—products such as amino acids and proteins, vitamins and hormones, and even small quantities of fats.

During the entire complex process, which may be likened to an automated chemical factory that goes on and on as long as raw materials are supplied and end products removed, many minerals are absorbed by the plant's root system for ultimate inclusion in chemical end products. Examples of such minerals are compounds of nitrogen, phosphorus, potassium, calcium, magnesium, iron, manganese, boron, and several other elements needed in such minute quantities that they are seldom deficient in the environment.

During millions of years of evolution, plants have shown a clear capacity for adapting to different habitats which may be deficient in one or more of the basic requirements. For example, certain plants are able to grow and function very well in the deep shade of mature forests, where sunlight is quite reduced, because their photosynthetic apparatus functions more efficiently than that of other plants. Not only do such shade-loving plants thrive on the forest floor, but they often do poorly if exposed to full sunlight. As another example, in dry desert regions all over the world plants have adapted to very low water levels throughout much of the year. This is accomplished by various developments: deeper root systems, chemical monitoring of the spacing of plants by which no one plant absorbs too much water from a unit of earth, reduced rates of water loss by changes in leaf chemistry and shape, or the adoption of a life cycle that fits into the brief period of annual downpour when the desert blooms and plants are able to grow and reproduce very rapidly before the land dries out again. There are many more examples of adaptation to what for most plants might be termed difficult environments.

The kinds of deficient habitats that will mainly concern us throughout this book are the usually acid, mineral-poor bogs and swamps, and the freshwater marshes and savannahs. It is in such locations that plants that have adopted carnivory may be found.

Anyone who sees a freshwater bog or swamp is impressed by the great variety of plant life—including many mosses, ferns, and orchids—and by the seemingly rich, black ooze of the wet soil in which the plants grow. But accurate chemical analysis of the soil in such areas frequently reveals that this initial impression is partly erroneous. First of all, the coffee-brown waters are very acid, and acid water along with frequent drenching rains quite efficiently leaches out many irreplaceable minerals. Second, in warm climates there is a very high rate of bacterial and other microorganic activity which in itself uses up the sparse supply of minerals that are needed by the larger plants. In cool climates such decay is delayed, but then the undecayed remains of dead plants and animals keep the much-needed minerals locked up. Third, a close examination in most cases confirms that the black ooze is little more than fine white sand along with a great deal of chemically sterile carbon or charcoal-like material, or the latter without sand. It seems that in such mineral-poor habitats there must have been considerable adaptation by all the plants in order for them to grow and reproduce.

To adapt to these environments deficient in many minerals and possibly in some cases to overcome the inability of their root systems to absorb required minerals, some plants retained the evolved capacity to trap and digest small animals. From the trapped animals, which are largely insects, the plants absorb what they need. As a result of selective adaptation, the plants that were able to trap animals survived and produced offspring with the same genetic characteristics.

The acceptable word "prey" is usually used in reference to the entrapped animals, but it is not literally correct in that the plants do not actively stalk and capture food in the sense that many animals do. Rather, the plant is able to lure or take advantage of incidental nourishing visitors by means of one of four kinds of trapping mechanisms, which will be discussed below.

After entrapment, the prey undergoes digestion. From a chemical standpoint the digestive process is quite similar in many respects to digestion in animals. Also, various microorganisms such as bacteria aid many carnivorous plant species in breaking down the prey into simpler, absorbable substances.

In the decades immediately before and after the turn of the century, many experiments were contrived to prove that digestive activity actually occurs in plant traps and to measure and define the nature of that activity. Some of these experiments were quite elaborate and their results still stand. For digestion to occur, certain enzymes must be present. Enzymes participate in the chemical reactions of biological organisms by causing the reactions to be completed rapidly at temperatures suitable for the maintenance of life. These reactions include the synthesis of more complex compounds as well as the reduction that occurs in digestion. The results of many experiments indicate that enzymes are responsible for digestion in the traps of carnivorous plants.

The next question concerns the source or sources of these enzymes: Did they originate from the plants themselves, secreted into the trap along with fluid as a response to entrapped prey of a suitable nature; or were they simply products of bacteria or fungi inhabiting the decaying detritus accumulated in some open traps? Like most questions of this nature in science, a categorical "either–or" answer is impossible, and it would be misleading to attempt to give one. It has been shown that some species of carnivorous plants have a complex enzyme-secreting system in small, specialized plant glands associated with the trap. Others with similar glands secrete practically no enzyme under sterile experimental conditions where the contribution of any microorganism can be discounted. And some plant traps function with no glands at all. At this stage, the answers are far from complete. Some plants seem to rely almost exclusively on their own enzymes, some seem to depend almost totally on bacterial action, and others take advantage of both sources.

Another question concerns what digestive products are actually absorbed by the trap of the carnivorous plant, which of these are truly required by the plant, and which are just passively absorbed. A second, related question is whether all the useful materials absorbed by the plant are simple minerals which may be lacking in the plant's habitat, or whether some are more complex, synthesized materials needed because, as a result of evolutionary change, the plant has lost the capacity to produce them. The individually studied cases are few and far from complete, so again we can give only some partial answers—merely clues in a highly complex problem that involves more than curiosity about carnivorous plants and actually cuts across the whole problem of the nature of adaptation.

Of all the mineral elements mentioned previously, the one that green plants need most consistently and in the largest amount is nitrogen, followed by phosphorus and potassium in more variable quantities. Acid soils are also quite deficient in calcium. All these elements are retained by "sweet" or basic soils—thus the gardener adds lime (a calcium compound) to "sweeten" or enrich soil that is too acid to permit most plants to grow well. Much research has centered on the idea that nitrogen is the limiting factor, or element most needed by carnivorous plants for sustenance and growth in their deficient environment, probably because nitrogen has long been prominent in soil and plant chemistry. But insufficient work has been done to establish the exact role that some other minerals or combinations of minerals may play in plant carnivory. For example,

recent preliminary observations indicate that potassium levels in soils, plants, and prey greatly influence the amount and rate of nitrogen absorption by carnivorous plants.

Very early attempts were made to find out whether anything could actually be absorbed by carnivorous plants. Researchers utilized harmless dyes which could be followed visually in their course through the plants. The air surfaces of most plants are covered by a thick, waxy layer called cuticle. The absorption of watery materials through a waxy layer of cuticle varies from slow to impossible. It was noted quite early that the absorbent interior surfaces of the traps of carnivorous plants lack cuticle. It therefore was possible to follow the dyes visually in their course through the plants. These were important preliminary results. Of course the experiments had very severe limitations.

Later, with the advent of radioisotope tracers wherein various portions of a material can be tagged with radioactivity and followed through the plant and in actual chemical changes in plant tissue, it was possible to conclude that absorption of certain materials did take place and that these materials were actually used by the plant tissues—that is, the substances did not just passively enter the plant tissues. So far these studies have been limited to nitrogen compounds, and we have only the published reports of studies by one worker using one species of carnivorous plants out of the forty or so on this continent alone; but it is a beginning, and it is certainly indicative that carnivory must be of some benefit to the plant.

Additional work on more general levels suggests that some carnivorous plants can subsist without trapped and digested animals, or that minute quantities of suitable fertilizers can be substituted by applying them to the roots, the trap interiors, or even the external leaf surfaces. However, a common observation in such experiments is that the plants are not as vigorous as in nature: they grow more slowly and do not become as large; they are more prone to disease; and very importantly, they do not reproduce as well, as is indicated by the production of fewer flowers and seeds, a reduced rate of seed maturation, and less rhizome budding.

So far, we have looked at carnivory from the viewpoint of an isolated, experimental plant. But plants occur in nature with other similar and dissimilar plants, with animals, and with an inanimate environment as parts of a community. There results a complex interaction of so many factors that one is awed and baffled in one's first attempts to picture the situation in perspective. The picture is further complicated by the fact that biological communities are not static; they are always varying and responding to assault and change. When a prime environment for carnivorous plants changes from wetland to grassy field, scrub, or forest as a result of natural or man-made activities, carnivorous plants and many of their wetland companion species disappear somewhat promptly, often in a rather specific order. They are apparently crowded out by forms more vigorous and better adapted to what has become essentially a new environment. It seems that carnivorous plants require the poor soils of an acid wetland to be competitive, soils where many other plants that under different conditions would be strong competitors cannot grow. When dryland plants that demand richer soils are finally able to spread into a reduced bog or marsh, carnivorous plants become the disadvantaged forms and disappear.

This is not so difficult to understand or accept in broad terms if viewed from a simplified but largely valid evolutionary angle. We began this section by mentioning the adaptation of plants to differing situations. Not all plants able to grow in the environment of an acid, mineral-deficient wetland adopted carnivory. Evolution seldom narrows to one pathway or one structural adaptation to solve a problem. Variation and gradual migration are the keys to the continuation of some life forms in a continually changing environment.

CARNIVOROUS OR INSECTIVOROUS PLANTS?

I will not belabor the point as to whether these plants should be called "carnivorous plants" or "insectivorous plants," but I will mention it lest the reader become confused by the use of both terms in conversation or in other publications. When carnivorous plants were first noticed and studied, the most obvious prey was insects; hence the term *insectivorous plants*. Later, species with more varied appetites were found. Skeletons of small birds and amphibians were found in some traps, and aquatic plants trapped small water animals that were clearly not insects. Thus the term *carnivorous plants* was coined to be more general and inclusive, and more accurate. It is the preferred term and the one we shall use throughout this book.

KINDS OF TRAPS

The traps of carnivorous plants are modified leaves that in some cases are so changed and adapted to their function that they resemble only remotely leaves as most people picture them. For example, the tall, often decorative tubular pitcher leaves of species of *Sarracenia* are frequently thought by the uninitiated to be flowers, and in most cases the trap leaves are far more striking than the plants' true flowers. This case of mistaken identity is somewhat ironic since it is accepted that most true flowers are decoratively structured in order to attract insects or other animals as pollinators. The trap leaves are also attractive to insects, but for a different end.

There are four types of traps in seed-bearing carnivorous plants of our region, and I have further divided these into two main groups, active and passive. I would reiterate that "active" is used in a restricted sense, not as it might be used in connection with animals of prey. A classification of these trap forms along with examples is in outline form below and can be correlated with the accompanying photographs and drawings.

Active traps.—Those in which some rapid plant movement takes place as an integral part of the trapping process.

1. Closing traps.—These are often referred to erroneously as of the beartrap type. The trap is bivalved; that is, it has two similar halves connected by a midrib. The two halves close on each other and thus trap the prey. This type is represented in the western hemisphere by only one species, *Dionaea muscipula* (the Venus' flytrap).

2. Trapdoors.—These are aquatic traps, relatively minute, and are represented by the genus *Utricularia* (the bladderworts). The trap is somewhat bulbous, with a flaplike door over a small entrance at one end. The stimulation of sensitive external trigger hairs near the trap entrance results in the opening of the door and an inrush of water with the prey. Afterwards, the door closes again.

Passive traps.—Those in which rapid plant movement is not an integral part of the trapping process.

3. Pitfalls.—These are characteristic of the familiar pitcher plants of the genera *Sarracenia* and *Darlingtonia*. The leaves are tubular with various other modi-

Fig. 1–1. Dionaea muscipula, *the Venus' flytrap, with a trap of the closing type.*

Fig. 1–2. Utricularia gibba, *a bladderwort, with a trap of the trapdoor type. The plant is aquatic and the bulblike trap is only 2–3 mm.*

Fig. 1–3. Sarracenia purpurea, *a pitcher plant, with a tubular leaf trap of the pitfall type.*

fications. The prey is lured to the pitcher opening, enters or falls in, is unable to escape, and is digested. 4. "Flypaper" or adhesive traps.—These occur in *Drosera* (sundews), and *Pinguicula* (butterworts). Numerous sticky glands cover the upper leaf surfaces, and the small prey is immobilized by becoming mired down. After entrapment, the stalked glands of *Drosera* do often move slowly and there frequently is some slow leaf folding in some species, but this is part of the digestive rather than the entrapment process.

This brief outline is for orientation; details of various traps and their activities will be discussed in the ensuing chapters.

CARNIVOROUS PLANTS AROUND THE CONTINENT

We have noted that carnivorous plants occur mainly in acid, freshwater wetlands. As is the case with all generalizations, this one has an exception or two. *Drosophyllum luscitanicum,* a native of Portugal and parts of Morocco, which will therefore not concern us further in this volume, occurs in semiarid regions. Of concern to us is a pitcher plant, *Sarracenia purpurea,* which can occasionally be found in alkaline marl bogs of the northeast as well as in its usual home in acid bogs.

Now we are going to make a quick tour of major sections of North America and take an overview of broad areas where categories of carnivorous plants can be found in suitable locations within the various areas.

Fig. 1–3. Sarracenia purpurea, *a pitcher plant, with a tubular leaf trap of the pitfall type.*

Fig. 1–4. Drosera capillaris, *a sundew, with traps of the "flypaper," or adhesive, type. The entire flattened rosette is about 5 cm across. Note the numerous gland hairs with sticky secretions at their tips.*

In the eastern two-thirds of Canada and the north-eastern quadrant of the United States, carnivorous plants are most often found in the classic acid sphagnum bog with which even the weekend naturalist is likely familiar. A northern sphagnum bog is usually an ancient glacial lake that has matured into a bog by becoming partially filled with undecayed plant detritus. It is then overgrown by large masses of various species of *Sphagnum* and other mosses, all tending to produce a very acid growing medium. As the bog further matures, or ages, other small plants, followed by larger woody plants, gradually move in toward the center of the former lake until finally a northern forest results. Carnivorous and other bog plants are then crowded out.

But while the bog is relatively young, one can often find the pitcher plant *Sarracenia purpurea* growing in profusion along with various sundews (*Drosera*) and some bladderworts (*Utricularia*), the latter either in the sphagnum or in the open acid water often found in the center of a bog. Many bogs are still basically large lakes or ponds with more open water than sphagnum mat, and in such areas the carnivorous plants grow along the lake margins. In more sandy, open places along the shores of large lakes and the Great Lakes, butterworts (*Pinguicula*) may be found.

We mentioned briefly the marl bog. This is a special area in which the seepage of spring water over a flat surface causes calcium carbonate to percolate up from limestone deposits. The alkaline marl results in conditions just the opposite from those of the sphagnum bog. But marl bogs do otherwise have some of the features that allow the growth of carnivorous plants, among them diminished nitrogen and other salts, constantly wet conditions, and the absence of many other plants that might become competitors. The presence of the normally acid-loving pitcher plant *Sarracenia*

purpurea in some of these areas indicates another form of adaptation that is not completely understood. A sundew, *Drosera linearis*, is adapted to marl bogs around the Great Lakes, where it is found almost exclusively. On the other hand, many other carnivorous plants will not colonize marl bogs. One final point on the ecology of the northeastern region is that it has been repeatedly glaciated, and after each ice flow retreated, plants have moved north again to repopulate suitable sites. Thus, plant populations in this region have been stable for relatively short geologic periods.

Progressing a little south, we come to the remarkable New Jersey Pine Barrens, which is a gross misnomer since to the eye of a naturalist they are anything but barren. But the early colonists did not find the broad, sweeping, sandy hills conducive to farming, so they declared them barren, and only timber and mining interests were able to utilize the region to any extent. Here there are many acid bogs along and in old lakes, slow streams, and sluggish springheads. The pitcher plant *Sarracenia purpurea* is quite abundant, and the kinds of sundews (*Drosera*) and bladderworts (*Utricularia*) become much more diverse; but butterworts (*Pinguicula*) are absent.

In the southern Appalachian Mountain chain from Pennsylvania south to its terminus in Alabama, there are occasional relic bogs that have survived ancient geologic activity that created these mountains from a peneplain (an almost level plain). The bogs are very much like the acid sphagnum type of the far north in general appearance, but they are most often found at a confluence of springheads or beside a stream rather than around the edges of maturing glacial lakes, which are not present in these areas. The kinds of carnivorous plants found in these mountain bogs are limited: the pitcher plant *Sarracenia purpurea*, a sundew (*Drosera*), a few bladderworts (*Utricularia*), and two other pitcher

plants that are unusual cases and will be discussed in Chapter 3. More bogs are found as the mountains recede into the eastern foothills and piedmont sections of the southeastern states, and the diversity of carnivorous plants increases as we approach the coast.

The last general area to consider in the east is the southeastern coastal plain, which runs as a great arc from eastern Virginia south and west to eastern Texas, including all of Florida. This area was suboceanic before the coastal uplift, and it is probable that the rich carnivorous plant life there is ultimately descended from plants of the former peneplain which has now been replaced by mountains and piedmont. The plants apparently migrated down rivers to habitats more similar to their ancestral locations. Since then, further cross migrations and evolution have undoubtedly occurred. The few forms adaptable to mountain climates were able to stay behind and evolve still further, some probably not adapting at all to coastal habitats, and some adapting to both mountains and plain.

The southeastern coastal plain is our richest area for both the number of species and the total population of carnivorous plants; about ninety per cent of the species to be discussed in this book can be found there. Many, such as the Venus' flytrap (*Dionaea muscipula*), are found there exclusively. While sphagnum bogs of the streamside or springhead type are found in this

---

Fig. 1–5. *A typical Appalachian Mountain bog. Note the background trees and mountaintop. The bog is grassy with a ground layer of sphagnum moss. There is water 2–3 cm deep in most places. This is a confluent spring bog, and the drainage stream is seen in the lower right-hand corner.*

Fig. 1–6. *A southeastern coastal plain savannah. The trees are lightly scattered among grasses and sedges. Some pitchers of* Sarracenia flava *can be seen above the grass in the foreground.*

region in abundance, the most characteristic habitat is a savannah, or grass-sedge bog. This is a low, flat or slightly sloping, sandy area with high water table and supporting predominant stands of grasses, sedges, and rather widely spaced longleaf pines. A healthy savannah is quite moist and acid.

Traveling rapidly across the continent, there is a paucity of carnivorous plants in the mid-plains and prairie: one species of sundew (*Drosera*) in wet pockets of the southern plains, and some bladderworts (*Utricularia*) in scattered aquatic sites. The deserts and eastern Rocky Mountains are devoid of carnivorous plants.

Parts of the Pacific mountain slopes are a different matter, particularly from northern California into Oregon. Again there are sphagnum bogs alongside or heading mountain streams, as there are in the eastern mountains. Curiously, there are several of the same kinds of sundew (*Drosera*) and bladderworts (*Utricularia*) that are found in the east, and a butterwort (*Pinguicula*) can also be found there. Quite unique is the California pitcher plant, *Darlingtonia californica*, a member of the same family (Sarraceniaceae) as the eastern pitcher plants but ranked in its own genus.

THE FUTURE OF CARNIVOROUS PLANTS

This section title propounds a vital question to which we can offer only some guesses, but they are largely well-founded guesses. Unfortunately, the outlook appears quite grim for many species. A number of factors contribute to this opinion.

Primarily, the most dangerous hazards are the result of man's modification of the environment for personal and often shortsighted ends. These modifications include, particularly, the control of fire and water levels in wetlands.

We will first consider the value of fire. The suggestion that there is any "value" in fire would seem anathema to those who in the past century have put all the efforts and sloganeering into the control of fire throughout our wild lands. But what we will be talking about is a specific kind of fire in certain specific areas.

Fire is necessary for the health of a bog. Bog core samples show that in ancient times there were many fires, as is indicated by charcoal layers and evidence of post-fire regrowth. Modern research in which bogs have been regularly fired over a period of years shows that a fast surface fire tends to remove detritus, competing herbs, and young woody plants that invade the margins of a bog as it goes through its natural maturation process (eutrophication) toward becoming forest land in the north and a scrub bog in the south. Periodic autumn firing, properly controlled, can greatly prolong the life of a bog. However, man the farmer, lumberman, and developer has either caused superheated holocausts that destroy everything in huge tracts of natural lands or has tried to control all fire. Many fine areas formerly inhabited by carnivorous plants have been absorbed into forest or scrub during the lifetimes of living botanists who have witnessed the process.

Second, in order to be able to approach timberlands in swamps, as well as to extend agricultural areas, many wetlands have been and still are being drained. Others have been converted to ponds or lakes. In the relatively flat terrain of the southeastern coastal plains, where an area of several hundred acres may not have a variance in net elevation of more than a few feet, drainage is easily accomplished and can be clearly seen in the newly created patches of seasonal desert amid an extensive network of roadside ditches throughout this section of the country.

To these two main factors can be added such secondary insults as pollution with fertilizer and toxic materials, willful vandalism, and the collection of plants by casual enthusiasts who are passing through. There have even been recent documented instances of the collection by commercial nurseries of entire stands of extremely rare forms. These problems do not eclipse the more basic situation of a radical change in the habitat, but they are not at all minor, the excesses of human nature being what they are.

So much for the grim side of the picture. On the other hand, there are ongoing attempts to preserve representative areas from the fate of neighboring locations. These efforts are having varying success. More people are becoming seriously concerned about the misappropriation of our resources, even those resources in which no immediate economic value is apparent. Nature conservancies, provincial, state, and national parks and a few local ones, local private groups preserving a small bog, and botanical gardens featuring native plants are all making some headway in setting aside, protecting, and managing wisely areas that include carnivorous plants. Some states have passed sweeping plant protection laws, although enforcement of these laws is difficult and at best erratic. Some commercial lumber companies in the southeastern United States have, on the recommendation of experienced botanists, taken upon themselves the task of sparing and even annually burning certain botanically valuable tracts of land which could have been devoted to tree farming. However, these same companies grant "carnivorous plant collection and sales rights" to commercial nurseries.

There is a lot to be done, and the situation is rather urgent. No individual is going to be able to make dramatic changes of any sort, and much of the damage is irreversible, short of reclaiming sites through radical treatment and then making massive transplantings. But individuals can participate in and support conservation groups which are trying to set aside a few extant representative areas, whether they are areas of park proportions or small bogs located on local farmland which might be purchased and properly maintained. On an individual basis, one can pursue one's citizen's rights by electing sympathetic legislative representatives or influencing the votes of those already elected, keeping in mind the realistic fact that a certain proportion of desirable natural lands must and will yield to basic economic and human necessity. Also on an individual basis, one can discourage vandalism and suppress one's own inclination to dig and try growing unusual native plants alongside the tomatoes and petunias. They will always die with such casual treatment.

The serious student can be of further help by assisting in preserving in artificial or barren natural bogs, in tubs, or in greenhouses, many species of carnivorous plants that are collected from condemned sites or purchased through reputable dealers. Dealers are supposed to propagate their stock rather than collect from the field to fill orders. A few reliable commercial sources for carnivorous plants are given at the conclusion of this book, along with hints on growing the plants successfully.

The last two sections of this chapter are intended for the reader who is not widely experienced botanically. They are very brief reviews of flower structure and function and of the system of Latin biological names. Those desiring additional information should consult any elementary botany text or some of the references mentioned in the final chapter.

FLOWER STRUCTURE

The flowers of most carnivorous plants take second place in attention to the trap leaves. But in many bladderworts (*Utricularia*), the flowers may be the most noticeable part of the plant and the part most easily used for identification. Flowers are, of course, important reproductive organs, and some familiarity with their general plan is necessary in any botanical study.

Reproduction is obviously fundamental to all living organisms. The usual results of wear and tear, age, and diseases would soon become self-evident if plants did not continually replace their losses. We may broadly divide reproduction into (1) *sexual*, the exchange of living material between two separate organisms of the same kind so as to diversify and enhance the genetic base of the species, and (2) *asexual*, the division of parent plants or parts of plants, which allows rapid local reproduction under sometimes adverse circumstances. Asexual methods of reproducing include the familiar bulbs, rhizomes, stolons, and various types of budding. Many plants utilize both general categories of reproduction. The flower is the organ of sexual reproduction in green seed-bearing plants.

Flower Parts

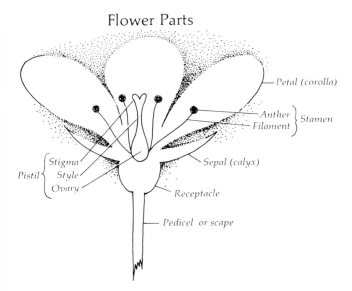

Petal (corolla)
Anther
Filament
} Stamen
Pistil { Stigma
Style
Ovary
Sepal (calyx)
Receptacle
Pedicel or scape

Most flowers consist of four basic whorls or layers of structures which, from bottom to top (or outside in), are (1) *sepals*, collectively called the *calyx*, which are usually green and leaflike but can be colorful; (2) *petals*, collectively called the *corolla*, which are usually quite colorful and often give the flower the general appearance by which it is recognized; (3) *stamens*, each consisting of a simple filament supporting a knoblike body called the *anther*, which actually produces the pollen, a fine dusty material; and (4) the *pistil*, at the top of which are one or more *stigmas* having sticky, sugary secretions that hold the pollen grains and support their germination. Supporting the stigma is the filamentous *style*, below which is located the enlarged, often bulbous *ovary*, which houses the egg cells and will ultimately become the seed capsule.

Some flowers have eliminated the calyx, the corolla, or both, and some species have either stamens or pistils but not both. None of these conditions occur in the North American carnivorous plants, however.

The entire flower is supported on a base called a *peduncle*, and the whole is atop a *scape*, or flower "stalk."

Sometimes, if there is more than one flower to a scape, another stemlike structure called a *pedicel* connects the flower base to the scape. The latter condition is especially prominent among bladderworts (*Utricularia*).

The sexual aspect of plant reproduction, the exchange of living material between two plants of the same kind or between compatible plants of two different but closely related kinds, ideally takes place when pollen from one plant is deposited on the stigma of another. This process is called cross-pollination. Pollination may take place with the aid of gravity, wind, water, or visiting insects or other animals transporting pollen from one flower to another. Although cross-pollination is conducive to the greatest benefits of sexual reproduction, many plants are capable of self-pollination if crossing cannot occur. While sexual in a partial sense, self-pollination, with respect to genetic exchange, is definitely inferior to crossing. Selfing, however, is still superior to asexual types of reproduction, since certain biological processes occurring in pollination and seed formation still allow some measure of genetic variation and recombination which cannot be achieved in vegetative reproduction.

If the deposited pollen grain has found a suitable medium in the sticky, sugary secretions of the stigma, it will germinate somewhat like a tiny seed. A pollen tube carrying one or more pollen nuclei actually grows down through the supporting style into the ovary, where a pollen nucleus unites with an egg-cell nucleus. This process is called fertilization. A complex series of microscopic cellular divisions takes place in order to balance the nuclear genetic material, and an embryo is formed. Around the embryo, nutritional material often develops along with a seed coat, and the ovary becomes a maturing seed capsule which opens when ripe. The seeds are then dispersed through various means.

One additional pair of definitions needs to be considered. In basic form, flowers may be of two kinds: (1) *actinomorphic*, or radially symmetrical. If the flowers are cut in half along any plane that passes through the center of the flower, two equal mirror-image halves will always result. In other words, these are the "perfect circle" flowers, such as day lilies, magnolias, pinks, and, among carnivorous plants, the Venus' flytrap (*Dionaea*), the pitcher plants (*Sarracenia*), and the sundews (*Drosera*). (2) *zygomorphic*, or bilaterally symmetrical. These flowers can be cut in only one plane through the center in order to obtain two equal mirror-image halves. Examples of such flowers are snapdrag-ons, peas, most orchids, and, among carnivorous plants, the bladderworts (*Utricularia*) and the butterworts (*Pinguicula*).

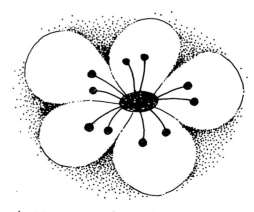

Actinomorphic Flower Form
Radially Symmetric

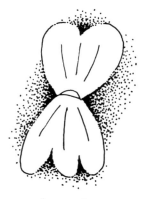

Zygomorphic Flower Form
Bilaterally Symmetric

HOW PLANTS ARE NAMED

The proper naming of plants has a long, complex, and interesting history, which we will not go into here. We will give only the most important principles of introductory nomenclature so that the reader can find his way in this or any basic biology book. We hope to overcome the initial reaction of recoil at the sight or sound of Latinized biological names.

The Latin names that botanists use represent a rather successful attempt to bring about uniformity in naming plants and to be certain that the same name is not used for two or more plants. Lately, they tend also to reflect theories as to various relationships between plants with respect to evolution and function. Latin names take some getting used to, but once one begins to understand them and to associate them with real entities, they become easier to deal with and their advantages over common names are appreciated. They are certainly here to stay; they are enjoying more common use among nonprofessionals, and they are the only names that are accepted officially.

Common or local "English" names suffer from a lack of specificity, accuracy, and uniformity because of local history and custom. Thus, the common name "pitcher plant" is incomplete in that it could refer to any of eight different kinds of plants or to all eight as a unit, even though the speaker may have a specific plant in mind. In another locality, the same pitcher plant could be called a yellow trumpet, a flycatcher, or a lily. In still other areas, the term "flycatcher" could refer to another of the several kinds of pitcher plants, or even to another class of carnivorous plants altogether. In other words, common names frequently do not have universal application.

The keys to eliminating this confusion revolve around two factors: the standardized basic Latinized binomial system, which has been in use for three hundred years, and the results of several recent international congresses of botanists, who decided how the system should be used and when and how it should be modified. The resulting rules are far from perfect, but one can establish official names—one for each kind of plant—and these names will be used and recognized by legitimate botanists the world over. The same name cannot be used officially for any kind of plant other than that for which it was intended. There results a common reference base and a classification that sweeps aside much of the confusion that would be involved in using common names alone.

The basic Latinized name of any plant consists of two parts or words: the genus name first, followed by the species name. The genus name is somewhat analogous to the human surname, and the species name can be compared to the "given" name identifying an individual in the human family. (Actually, the species name identifies a whole population of like plants.) All the members, or species, of a genus are similar to each other in some ways, as is true in human family resemblances, yet they are sufficiently different to warrant their species names. As a group, one genus is quite different from another, but even among an assortment of genera (plural of genus) there may be some basic similarities and some groups of genera thus can be blocked into botanical families. There are even higher divisions, right up to that between the animal and plant kingdoms, but they need not concern us at this time.

As an example of classification by family, genus, and species, let us mention the pitcher plants of eastern North America. They are members of the family Sarraceniaceae, which includes three genera of plants from North and South America. The genus name for the eastern North American pitcher plant is *Sarracenia*, which in this case as in many is very similar to the family name since it is felt to be the prime example (or type genus) of the family. Within the genus *Sarracenia* there are eight widely recognized species, including *Sarracenia flava, Sarracenia purpurea, Sarracenia alata,* etc. Thus the botanist can speak of a family by its proper name when discussing a very broad group of somewhat similar plants, a genus when he wishes to mention more closely related plants as a subgroup within that family, and use a binomial name (often called just "species" in jargon) when discussing a single kind of plant.

We must now mention a few simple rules as established by the botanical congresses devoted to this problem. The name of a genus or species is ideally a Latinized description of a distinctive feature of the plant, but it can be derived from another source such as a person's name or a geographical location. The genus name always begins with a capital letter and the species name with a small letter, except when the species is named after a person, in which case a capital letter *may* be used, but the trend is away from all capitalization in species names. A binomial name is always printed in italics or underlined when handwritten or typed.

In very formal botanical writing, the binomial name is followed in ordinary roman type by the name of the person who first named the plant, or by several names in some cases when the name has been properly changed over a period of time. Most often, a standardized abbreviation of the man's name is used if he is well known among botanists.

A genus name can be used only once in botany, and no two kinds of plants within a genus can bear the same species name. As an example of the latter, our *Sarracenia* genus of pitcher plants can have only one species named *Sarracenia purpurea*, but the species name *purpurea* can be used in other plant genera, as in the case of *Utricularia purpurea*. Following the rules, there is no *Sarracenia* other than one of the eastern North American pitcher plants.

Another rule is that certain abbreviations are allowable. If one is discussing *Sarracenia* in a writing, one can abbreviate the genus to the first letter after once using the full genus name. So, if I mention *Sarracenia flava* in a chapter or paragraph and have not during the course of the discussion mentioned another genus beginning with the letter *S*, I may abbreviate the name to *S. flava* the next time I use it.

Not all is settled by any means in plant naming and classification (the science *and* art of taxonomy), and we will have to confront a few controversies in this book. There is often debate about which name has legal priority in a case where two botanists have accidentally or willfully given the same kind of plant two different Latin names. There is much serious and legitimate discussion about what actually constitutes a species—where one draws the line between "kinds" of plants. The genus *Sarracenia* has eight commonly accepted species, but many serious and learned botanists with good arguments would declare that there are as many as ten species. However, in order to get additional species established they would have to go through the procedures that have been prescribed for such changes, and all botanists would have access to the information.

The complete classification of a whole group of plants, or a single species, can be changed if certain formal steps with respect to studies and the publication of those studies are carried out, and then—most importantly—if enough botanists agree with those studies and use the new classification. Whether they do or not, all botanists would know precisely what was being discussed if they came across a paper on the subject. Admittedly, as in the case of rules and laws everywhere, a considerable amount of nonformalized agreement is necessary to make the system work well.

Another area of controversy in taxonomy is the use of subdivisions finer than species, these being, in order of specificity, subspecies, forms, varieties, and races. A few subspecies, representing differences in large populations of a species that occur over wide areas or with discontinuous ranges, are formally but reluctantly recognized by many students. Forms and varieties are more or less tentative and most often represent populations of more importance to the ecologist and the evolutionist than to the taxonomist. Often they are merely horticultural terms, many of which are eventually dropped altogether, but the plants they represent may after further study be elevated in stature to subspecies or species—although this happens rarely. Many taxonomists claim that these finer subdivisions are artificial, a product of man's propensity for organization rather than representing actual biological groupings. Others claim that the variants they represent are the stepping-stones of evolutionary change. At the very least, their value lies in their usefulness in communication, if nothing else. The horticulturist can find a varietal name very helpful for accurate and brief reference when communicating with other horticulturists, and the ecologist or evolutionist studying the derivation and relationship of plants can find the subdivisions equally useful for quick and pointed reference to natural variations that he may recognize for one reason or another.

Proper abbreviations for handling these subdivisions are ssp. (subspecies), f. (forma), and var. or v. (variety). The subdivisional name is italicized, but the reference abbreviation is not. Examples are *Sarracenia purpurea* ssp. *venosa* and *Drosera filiformis* f. *tracyi*. Often the divisional abbreviation is omitted, as in *Sarracenia purpurea venosa*. We will, as we go along, mention instances of such subdivisions among carnivorous plants and some of the pertinent arguments pro or con.

II. The Venus' Flytrap (*Dionaea muscipula*)

BOTANICAL NAME: *Dionaea muscipula* Ellis ex L. A monotypic genus; that is, there is only one species in the genus. Family Droseraceae.

COMMON NAMES: Venus' flytrap, flycatcher, tipitiwitchet, catch-fly sensitive. (The latter two names are ancient.)

RANGE: Quite localized in scattered savannahs of southeastern North Carolina and neighboring eastern South Carolina in an approximate landward radius of 60–75 miles around Wilmington, N.C.

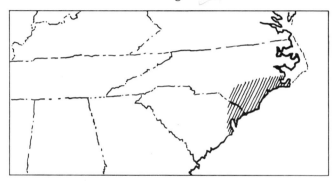

Fig. 2–1. *A plant of* Dionaea muscipula *as typically seen in the field.*

FLOWERING SEASON: Late May through early June.

TRAP SEASON: Some plants remain evergreen in protected situations, while many die back for the winter. New traps begin growing in March and continue into October.

DESCRIPTION.—The plant is a rosette of leaves that radiate out from a central point, the leaves being totally or partially reclining. The rosette measures 10–14 cm across when mature. The leaves arise from a somewhat elongate, fleshy, white rhizome (underground stem) often miscalled a "bulb." The rhizome elongates and enlarges annually. Fibrous roots descend 10–15 cm. The green leaves grow up to 12 cm long. They are of two parts: a narrow to relatively broad leaflike petiole (leaf stem) nearer the rosette center, and a leaf blade modified into a unique trap, measuring up to 3 cm long.

The flowers are on a 1–30 cm scape. They are actinomorphic, have white petals, and are about 1.0–1.5 cm across. After fertilization, tiny, black, pear-shaped seeds set (mature) in 6–8 weeks. These will germinate immediately when sown on a suitable substrate; storage at warm temperatures results in a lower rate of germination. The plants mature from seedlings to flowering age in 3–4 years, and estimates of the age of the oldest known living plants are nearly 25 years.

The traps consist of two clamshell-like halves. Around the free margins (unattached edges) are numerous stout guard hairs and minute nectar glands. The trap is normally in a 45°–60° open position when undisturbed. The interior of each half is lined by nearly microscopic digestive glands, which give the surface a finely granular or cobblestoned appearance. Each inner half has also three smaller, finer trigger hairs in a triangular pattern (rarely anomalous plants have up to six hairs per half) which, when properly stimulated, initiate trap closure. The trap lining is colored variably green to pale yellow to bright red with frequent intermediate shades and patterns noted.

The usually bright coloration and the secretions of sweet nectar by the marginal glands may attract prey to the interior of the trap, where the insect brushes against one or more of the trigger hairs. Initially, trap closure is quite rapid until the guard hairs mesh, effectively incarcerating the small prey in a barred sarcophagus. The slower, secondary phase of closure results in the margins sealing tightly together so that the whole trap becomes a flattened, stomachlike pouch. At this stage, the margins of the trap halves evert slightly. If live prey—not a raindrop or a piece of windblown debris—has been caught, digestive fluids are then secreted into the interior of the closed trap. Apparently, the struggles of the prey and certain

Fig. 2–2. *A plant of more upright habit. Note the numerous red-lined traps with prominent marginal hairs.*

Fig. 2–3. *Close view of two traps, one with previously digested insect remains, the other after the rapid phase of closure. The intermeshing guard hairs hold the prey in until the slower closing phase is completed.*

17 / *The Venus' Flytrap*

chemical compounds that emanate from the prey (such as amino acids) stimulate more copious secretions of digestive fluid with greater concentrations of enzymes. Digestion then occurs, and the nutrients are absorbed at the bases of the glands over a period of 3–5 days, depending on the temperature, the size of the prey, its nutritive value, etc. Afterwards, the trap reopens. The dry, chitinous remains of the insect stay in the trap or drop out.

Each trap may be stimulated mechanically by touch to close about ten times before it will no longer respond. After such closures, the trap reopens the next day, since there is no animate matter to digest. If the prey is not too large, each trap may catch and digest up to three times, after which it ceases to function. Very large catches result in the death of the trap leaf, but new ones grow more or less continuously all season.

In order for the trap to close, any one of the trigger hairs on the inner surface must usually be touched twice, or any two hairs must be touched once each in succession. At temperatures above 40°C., however, one stimulus suffices in half the cases. The longer the period between the two stimuli, the slower the closure. The very quick reaction no longer occurs if the period exceeds the range of 20–40 seconds. If the stimuli are farther apart, closure will eventually occur, but it is extremely slow and multiple stimuli are required during closure to complete the process. It is felt that the struggle of the live prey inside the trap continuously stimulates the closure mechanism and the secretion of

Fig. 2–4. *Interior of trap. Note the three triangularly spaced trigger hairs on each inner surface of the trap.*

Fig. 2–5. *Appearance of trap after catching a live insect, the slow phase of closure having been completed. The edges are tightly sealed.*

the digestive enzymes until the very tight, pouchlike seal is formed.

The mechanism of closure is not entirely understood. A small but consistently measurable action potential, or electric current, crosses the leaf after the trigger hair is stimulated. Sometimes, the stimulation of parts of the leaf other than the trigger hairs initiates the action potential and results in closure. The fact that the total number of closures for any one trap is limited indicates that some cell growth function, metabolic process, or both, is ultimately exhausted. The fact that the number of repeat closures is more limited if prey is caught each time certainly indicates that the process is accumulative rather than exhaustive; that is, that possibly the storage or metabolic manufacture of some product—perhaps starch or protein—may cause the process of closure to become inhibited sooner than it would with inanimate stimulation.

In spite of initial appearances, the trap does not close like a bear trap. The two halves do not rotate on the midrib like a hinge. When open, the outer surface of each trap half is concave, or dished in, while the inner surfaces are bulging inwards. During closure, these surface conformations are reversed, so that the free edges are quite suddenly brought closer together— close enough for the strong guard hairs to intermesh.

Fig. 2–6. *A cluster of plants in flower. The scapes are disproportionately long for the size of the plant, perhaps so that the flowers are raised above the grasstops for effective pollination, and so that the potential pollinators are not more tempted by the traps farther down.*

Fig. 2–7. *Flowers of Dionaea muscipula.*

Fig. 2–8. *Typical view of Dionaea flowering in the field. The small white flowers are just visible above the grasstops, while the vegetative parts are partially obscured.*

19 / *The Venus' Flytrap*

This process can be noted by close, careful observation as well as by high-speed comparison photographs or motion pictures. However, the change in surface conformation explains only the rapid phase of closure. During the slower, sealing phase, no further significant change in surface conformation is observable except for the eversion of the very margins, and there is even some slight loss of the new convexity of the outside surfaces. At this stage, some hingelike rotation on the midrib likely occurs.

As we mentioned, the bladelike leaf petiole has a variable morphology. In shadier locations, or in the early spring prior to flowering, the petiole is quite thin and wide and later ones become thicker and narrower. But even late in the season and in full sun, some petioles of adjacent plants remain wide, while others become almost stemlike or triangular in cross section. Both variations can sometimes be found in plants in a single location. Some observers feel that these variations are inherent in different genetic strains of the species, and they have recognized at least four such variations, including the two extremes and two intermediates. Another interesting observation is that the narrower the petiole, the more erect the whole leaf tends to be.

The color of the trap lining may follow a similar pattern. Generally, growth in bright sunlight brings out the brightest red color. However, some plants growing right beside the red ones, in the same light and soil, remain green or yellowish or are even variegated red and green.

Anomalies of morphology aside from that of the petiole are also occasionally noted. We have mentioned that the trap can have up to six trigger hairs per half.

Fig. 2–9. *Seed, which ripen exposed.*

Fig. 2–10. *Seedlings.*

Another individual, nongenetic variation is a double trap on one petiole. An occasional flower anomaly is vegetative apomixis, in which the flower parts—sepals, petals, stamens, and pistil—are replaced by miniature plantlets which can be rooted and grown to normal plants. This occurs mainly when spring weather has been uneven during the early period of flower bud initiation, with cold nights alternating with warm, sunny days. The author produced this phenomenon some years ago in Ohio by growing the plants on a windowsill during March. The warm sunlight streaming through the window initiated early growth and flower budding, but the nights were so cold that the glass next to the sill was often frosted on the inside in the mornings. It is likely that other such environmental shocks, perhaps involving chemical substances such as one or more of the plant hormones, could precipitate the process. Finally, even the widened, bladelike petioles are capable of producing vegetative buds. The process has been utilized in culture in order to propagate the plants rapidly, and the phenomenon has occasionally been observed in nature.

GENERAL.—Largely because of habitat changes, *Dionaea muscipula* is markedly decreasing in numbers throughout a shrinking range which is none too large to begin with. Indeed, earlier reports document a far more extensive range in the Carolinas than we are able to report now. *Dionaea* does tend to remain on a deteriorating site longer than many associate carnivorous and noncarnivorous plants, particularly pitcher plants (*Sarracenia*). The plants grow in moist, sandy, acid savannahs among wire grasses, sedges, and many native orchids (*Pogonia, Calopogon, Platanthera, Spiranthes,* etc.) between rather widely spaced longleaf pine trees. *Dionaea* will tolerate short periods of drought and flooding, submerged plants having been observed catching small aquatic animals. Since the underground stem of the Venus' flytrap is well protected, *Dionaea* is among the first plants to sprout back strongly in a burned area. If the area is not burned from time to time (or if the water table drops), other herbs, shrubs, and trees encroach and quite quickly crowd out the smaller *Dionaea*, since the habitat is then a completely different one. Thus a rapid surface fire in the autumn is actually quite beneficial.

In late summer, because the neighboring grasses are quite tall by that time, considerable search is required by the uninitiated before he finds the plants, often after walking over them for some time! The best time to observe *Dionaea* is in early spring when grasses and sedges are shorter, and especially when the flowers, lifted by the tall scapes above the grass tops, can be seen easily even from an automobile.

Dionaea is rather hardy climatically. Outdoor experimental transplants have thrived as far north as New Jersey and some bogs in Pennsylvania. Incidentally, such transplants, which were often conducted without fanfare to discourage vandals, have nearly led to some embarrassing results when skilled naturalists, unaware of the experiments, have come upon the plantings during walks and nearly rushed to publication with the news of *Dionaea*'s supposed range extension.

While there is certainly much more to learn about all the plants in this book, *Dionaea muscipula* undoubtedly remains the pet among most students of carnivorous plants, botanists in general, and naturalists of all walks.

III. The Eastern North American Pitcher Plants (*Sarracenia* L.)

The Genus

BOTANICAL NAME: *Sarracenia* L. Named after Dr. M. S. Sarrazin of Quebec, an early discoverer. Family Sarraceniaceae.

COMMON NAME: Pitcher plant. (More specific common names will be mentioned with each species.)

RANGE: Generally, various species can be found in appropriate bogs, savannahs, and other wet places throughout eastern North America.

FLOWERING AND TRAP SEASONS: Flowering periods vary from March to June, depending on species and locale. Most pitchers die back over the winter, and new ones sprout with or soon after flowering.

DESCRIPTION.—Pitcher plants are all basically perennial rosettes of leaves modified into traps that arise from long rhizomes (underground stems) which have fibrous roots. The pitchers of various species range in height from 10 to 120 cm. The rhizomes often branch and have several growth crowns, so that an apparent clump of plants may really be a clone; that is, the plants may all be connected.

The oldest part of the rhizome is often dead. Two species regularly give rise also to flattened, moderately wide, elongate, leaflike structures that are probably modified petioles (leaf stems) and are known as phyllodia. These usually appear at the end of the growing season and are often referred to as "winter leaves," since they remain throughout dormancy. The phyllodia, which assume various shapes, can be helpful in determining species.

The prominent, often highly decorative trap leaves are tubular, appearing somewhat like elongated funnels or cornucopias. Located at the top is a lobe called a lid or hood. This is usually supported on a narrower column of varying prominence. The hood or lid may be reflected over the pitcher opening or may be, as it is in one species, vertical. The lid is immobile. Running down the axial seam (that which faces the center of the plant) of the pitcher leaf is another lobe which is flattened and winglike, its size and character varying with the species and growing conditions. This structure is known as an ala or wing. Pitchers may be colored from green to shades of red, yellow, or white. The pitchers assume two general habits according to species: either erect or nearly erect to totally decumbent.

The pitchers possess several interesting adaptations which serve to lure and entrap prey. The bright coloration of the pitcher and the secretions of nectar along the margins of the hood, the rolled lip opposite the column, and in some cases the free margin of the ala lure ground and flying prey to the pitcher opening. The inside of the lid is lined by stiff, downward-directed hairs, which encourage descent and discourage ascent. These hairs vary in size and effectiveness according to the species. Contrary to common belief, the lid does not snap down to close off the pitcher after the insect is caught. Its exact function is not totally clear, but it does, to a degree, prevent the contents of the pitcher from being diluted by rainwater in species where the

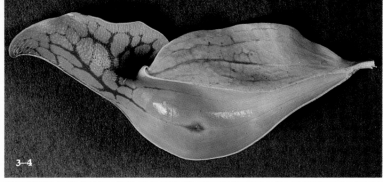

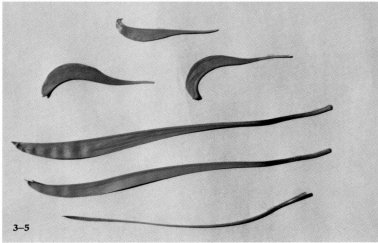

Fig. 3–1. Sarracenia oreophila. *An example of a pitcher plant with an erect habit.*

Fig. 3–2. S. purpurea *ssp.* venosa. *A common pitcher plant showing a decumbent habit.*

Fig. 3–3. S. purpurea *ssp.* venosa. *Notice the hood with dense, long, stiff hairs lining the inside and pointing downward.*

Fig. 3–4. *Longitudinal section of pitcher of* S. purpurea *ssp.* venosa, *showing four basic zones: the inner surface of the hood lined with stiff, downward-pointing hairs; a smooth waxy zone; a glandular and absorptive zone without cuticle; and a lower zone of intermeshing hairs.*

Fig. 3–5. *Comparison of phyllodia ("winter leaves"). Smaller, sharply curved* S. oreophila *at top; longer and straight* S. flava *below.*

lid is actually reflected over the pitcher opening. The lid must have a different function in *S. purpurea*, where it is vertical.

After having ventured to the brim of the pitcher mouth or the underside of the lid, the insect frequently overextends its footing and falls in. The interior of the upper one-third to one-half of the pitcher is lined with a smooth plant wax which impedes footing in most cases, and there is seldom enough room to begin flight. Deeper in the pitcher, waxy cuticle is absent, and the unwaxed surfaces are capable of absorbing digestive products. All species have a deeper, downward-directed intermesh of hairs which further helps prevent egress. There are digestive glands that may secrete protein-digesting enzymes and fluids of various kinds and concentrations. Those pitchers that do secrete enzymes have a small quantity of them in their fluids before entrapment occurs, and the concentration of enzymes tends to increase after initial digestion and absorption. The relative digestive efficiency of these fluids and enzymes from the plant and their possible concert with microbial activity is still being assessed. (See Chapter 1, pp. 2–3.)

The effectiveness of the trap varies according to habitus and species. All the erect traps are extremely efficient, often filling to the point that excess insects can freely walk or fly in or out. The widely flaring and reclining pitcher of *S. purpurea* seems less effective; in fact, it is theorized that this species actually drowns its victims.

The digestive mixtures of the pitcher are not universally effective. Various protozoa and insect larvae, for example, have adaptations to resist digestion, and in fact they breed in the pitcher. Actually, the contents of the pitcher comprises a complex little ecosystem of algae, fungi, bacteria, protozoa, other microbes, and various resistant insect larvae.

The flower structure of *Sarracenia* is generally the same for all species, the only variation being in size, odor, petal color, and some details of petal shape. The flowers appear in early spring, usually before or as new pitcher growth begins. The scapes are tall, to 70 cm in some species, and they support a single nodding flower. The unique floral structure serves to encourage cross-pollination, although experiments have shown that artificial self-pollination is quite successful in producing viable seeds. As the spherical flower bud approaches opening, the scape assumes the shape of a shepherd's hook, and the actinomorphic flower opens facing down. There is an unusual modification of the style in that the distal half is expanded, so that the whole looks like an opened, inverted umbrella and is commonly referred to as such. The umbrella has five points between which hang the pendulous, elongate petals, which are strap-shaped to obovate (rounded, but wider than long). At each umbrella point is a small, *V*-shaped cleft, at the lower point of which is located one very small stigmatic lobe. The rounded ovary at the base of the style and the numerous stamens are located inside a sort of floral compartment, at the top of which are the five sepals, three bracts, and the bases of the petals; the pendulous portions of the petals hang along the sides of the compartment like drapes, and beneath is the cupped, expanded umbrella. At the bases of the petals are nectar-secreting glands.

The insect pollinator, often a bee, is probably attracted by the color of the petals, the nectar, and the odor, and usually lands at the bases of the petals beneath the frequently reflexed sepals, where it circumnavigates the flower several times. It finally enters at the only visible narrow parting of the petal "curtains," over a point of the umbrella, and thereby brushes any pollen collected from previous flowers onto the stigma lobe. Inside the flower, pollen has been

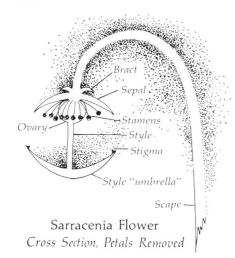

Bract
Sepal
Ovary
Stamens
Style
Stigma
Style "umbrella"
Scape

Sarracenia Flower
Cross Section, Petals Removed

3–6

3–7

shed from the overhanging stamens and has fallen to the floor of the cupped umbrella, where the bee may collect it, or it may collect it while investigating the stamens and nectar glands near the flower base. When the bee leaves the flower, it may not retrace its steps by passing over one of the umbrella tips and stigma lobes, since this offers poor footing prior to flight. Instead, the insect exits more often by pushing out one of the pendulous petals and flying from the wider umbrella edge between two points, thereby minimizing the chance of depositing newly collected pollen on the plant's own stigma.

After pollination and fertilization of the flower, the petals drop, and the flower frequently assumes a more nearly or totally erect position. The sepals and umbrella remain all season as the seed matures. Even if the flower has not been fertilized, the petals fall within two weeks of the flower's opening. In autumn, the brown, tubercular seedpods split at five seams and shed teardrop-shaped, 2 mm, light brown to pink-gray seeds. Dispersion is probably most often by gravity and water. The seeds require a period of stratification (damp winter chilling) before they will germinate. This characteristic effectively prevents the freezing of tender seedlings during the winter months.

GENERAL.—A bog or savannah of tall, golden-yellow *S. flava* or brilliant, white-topped *S. leucophylla* growing so thickly that one cannot walk without stepping over plants, is indeed an attractive and often startling

Fig. 3–6. *Flower of* Sarracenia flava, *typical of the genus. One petal has been removed to disclose the umbrella, the stigma point, and the stamens.*

Fig. 3–7. *Umbrella of* S. flava *flower spread out to show points and stigma lobes.*

sight. Equally interesting, on a "trot" through a northern sphagnum bog, is coming upon clumps of gaping pitcher mouths of *S. purpurea*, which are sometimes just above sphagnum level while the rest of the plant is nearly buried in live red-and-green moss. Pitcher plants are certainly our largest and among our most easily observed carnivorous plants, but unfortunately they are very susceptible to change in their habitat. Very few of the heavily populated, multi-acre stands described earlier in the century are still extant in the southeastern ranges. More often, one will find only scattered clusters or individuals, sometimes clutching the bank of a ditch that has drained a nearby savanna.

Drainage and fire prevention allow dryland plants to move in, and these then compete with *Sarracenia*. The natural late summer and fall fires of years past helped prevent such recolonization by clearing the site of debris and competing plants while the underground stems of the pitcher plants were protected. Many field experiments have demonstrated the beneficial effects of regular fires in a bog or savanna. There is some indication that fire may also release minerals tied up in dry, dead pitcher leaves and dead insects so that rains may leach them into the soil.

Regarding the effects of the control of competitors, I recall a situation in Georgia where a landowner had fenced off a section of savanna in which he was grazing cattle. He noted that the "lilies" (*S. flava*) were increasing inside his fence where the cows were grazing, while outside, the plants were decreasing. His puzzlement was further compounded by the failure of a yearly firing of the pasture to control the pitcher plants. Actually, he was burning competing plants, and the cattle were devouring all the grasses, herbs, and young woody plants—in other words, weeding. His misguided attempts to eliminate the pitcher plants in order to grow a good stand of pasture grass actually maintained a virtual garden of *Sarracenia*.

As we indicated, many small animals are capable of bypassing a pitcher or even converting it to their own use. Small spiders, snails, slugs, and frogs may visit the lips of the pitcher openings in search of food or

Fig. 3–8. *A dense stand of* S. flava *in Georgia.*

Fig. 3–9. *A multicrowned plant of* S. flava. *Note the short grass kept cropped by overgrazing cattle, which do not eat the pitcher plants.*

prey. The occasional report of frog skeletons in pitcher contents indicates that circumvention is not always successful.

The larva of a fly (appropriately named *Sarcophaga*) resists digestive action by the secretion of anti-enzymes. It feeds on the debris and pupates in the depths of the pitchers, doing no harm at all. People tell of opening "lilies" to remove the maggot and using it as fishing bait.

Within some pitchers, an unusual grass-cutting wasp builds a condominium of incubators consisting of alternate layers of dry grass and egg compartments with paralyzed crickets for the larvae to feed upon when they hatch.

But the greatest real attack on pitchers comes from a small yellow-and-black moth, *Exyra*. There are three species, each attacking different groups of species of *Sarracenia* in different ranges. The adult moth is able to walk about freely on the slippery, waxy surface of the inner pitcher where it hides during the day. If disturbed, it will back down the pitcher still further. If removed, it will promptly flutter to another pitcher and seek a new hiding place. The female lays one egg per pitcher, and the larva wreaks havoc.

The larva often spins a dense web across the pitcher mouth, closing off any further trapping. Then it feeds on the inner layer of pitcher tissue, causing the dried, papery top to fall over and seal the pitcher off from rain and interference. The pitcher becomes a private feeding and rearing area. The brown, collapsed tops of the pitchers are telltale signs of infestation, and many bogs are severely afflicted. In very severe climates, the larva may leave a mature pitcher and winter over for several seasons in the dead remains of previous years' growth. In the spring, it enters at the apex of a newly developing pitcher before it opens and girdles the top.

Fig. 3–10. *Pitcher of* S. flava *opened to show infestation by unusual, grass-cutting wasp* Isodontia. *Plugs of grass alternate with stunned insects for the larvae to feed on when the eggs hatch.*

Fig. 3–11. *Pitcher of* S. flava *infested with larva of* Exyra *moth. The top has collapsed because the inner layers of the wall have been consumed.*

Before it pupates, the larva cuts two holes in the lower portion of the pitcher: one above the accumulated waste frass, to be used for escape as an adult moth (moths do not have cutting mouth parts), and a lower hole for drainage in case some water seeps in during the wet winter season. But there is partial rebuttal by natural balances. Some birds have learned that the holes mean larval or pupal food within, and slash marks made by beaks indicate that many of these parasitic insects never reached adulthood.

Since many larvae of *Exyra* do overwinter in the fallen, brown pitchers of the previous year, fire is again helpful to the plant colony by burning old litter from the previous growing season and destroying the harmful insects. Indeed, the most infested stands of plants are frequently those protected from fire.

The whole field of associations between pitcher plants and all the components of their environment is a fascinating study with a great deal yet to be explored. I have never gone into the field without returning with some new perspective, question, or idea.

The Species

Sarracenia purpurea L.

BOTANICAL NAMES: *Sarracenia purpurea* ssp. *purpurea* Wherry; *Sarracenia purpurea* ssp. *pupurea* f. *heterophylla* (Eaton) Fernald; *Sarracenia purpurea* ssp. *venosa* Raf. These are three more or less recognized entities within the basic species.

COMMON NAMES: Northern pitcher plant, southern pitcher plant, sidesaddle plant, pitcher plant, huntsman's cap, frog's britches, dumbwatches. (The last is an interesting local term used in the New Jersey Pine Barrens, where the expanded style and sepals that remain after flowering were thought to look like open watches without hands, and therefore mute.)

RANGE: The species has a wide distribution as far west as northeastern British Columbia, where it was recently found, and over a good part of the eastern third of the United States and Canada, with an unexplained skip area in middle eastern Georgia.

FLOWERING SEASON: From as early as March in its far southern range to July or August in the north.

TRAP SEASON: Traps tend to be evergreen unless unduly exposed.

DESCRIPTION.—The pitchers are curved and decumbent, measure to 45 cm, and widen very prominently toward the mouth. There is a large, often slightly undulate ala. The hood rises vertically and is lined by long, stiff hairs. The edges of the hood are quite undulate and have lateral wings. The color of the pitchers varies from bright yellow-green to dark purple and is most commonly a middle variation with strong red venation.

Flower petals, sepals, and flower bracts are mainly rose pink to dark red. (See exceptions below.) The flower has a moderate odor of mixed nature, both feline or musty on one hand, and sweet on the other. Some have compared it to that of green peaches or peach twigs. The sweet component is most easily detected early in the morning or in shaded plants, while the feline odor becomes predominant in bright sun or as the day progresses.

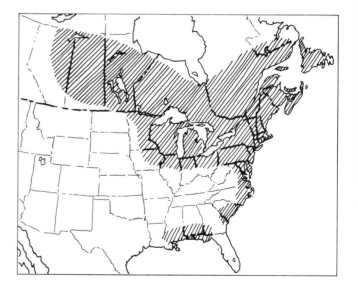

Since the species covers such a wide territory, it is logical that there should be some evolutionary variation on the periphery of the range. There is, and the variations have been variously interpreted as separate species, subspecies, forms, or of no real significance. The preponderance of evidence leads me to the following interpretations for the present: The plants of the northern extremity of the range should be designated *S. purpurea* ssp. *purpurea*, a noteworthy form being *S. purpurea* ssp. *purpurea* f. *heterophylla*; the southern plant should be designated *S. purpurea* ssp. *venosa*.

S. purpurea ssp. *purpurea* has narrower, longer pitchers, which are smooth and glabrous (hairless) on the exterior. The diameter of the hood when the wings are pinched together does not exceed the diameter of the pitcher. Venation is frequently present when plants are growing in full sunlight, and there may be a diffuse, coppery purple color to the upper pitcher. Wintering pitchers of the northern subspecies often turn deep maroon and become green again in the spring.

The form of the northern subspecies, *S. purpurea* ssp. *purpurea* f. *heterophylla*, is found in one county in

Fig. 3–12. *Two pitchers of* S. purpurea *ssp.* venosa *growing on the surface of a pond in the New Jersey Pine Barrens.*

Fig. 3–13. *Flower of* S. purpurea, *typical of a red-flowered* Sarracenia.

Michigan and in some eastern Canadian bogs in limited numbers, even though it may tend to dominate in an individual bog. This plant is without any red pigment at all, the pitchers and flower parts being yellow to yellow-green. Intermediates or form hybrids are easily found in such bogs if the typical subspecies is also present.

The southern plant is designated *S. purpurea* ssp. *venosa*. The pitchers are wider and stockier and have more prominent, coarse, red venation, always a fine coat of wooly hair (more easily felt than seen) on the exterior, and much more expanded hood wings.

This seems to be the best classification consistent with a broad perspective of our present knowledge. The two major subspecies are widely separated at the extremities of their ranges, and they have important biological differences and distinctive adaptive features that cannot be ignored.

When plants of one extremity of the range are transplanted to the other, they generally retain their characteristic features, but in a relative way; there is a tendency for the transplant to come to resemble superficially the native. Therefore, local soil, water, and climatic conditions do play some part in determining the plants' forms, but always within the governing genetic framework of any group of plants.

Where the ranges of the two subspecies merge in the New Jersey Pine Barrens, one can find both forms either in separate bogs or often side by side in the same bog. There are, of course, many interbred intermediates as well.

Parallel to these observations, the species is host to the larva of a harmless mosquito of the genus *Wyeomyia*, one species of mosquito reportedly inhabiting the pitcher fluid of the northern plant, and another species, that of the southern. It has been suggested that, where the two subspecies intermingle in the New

Fig. 3–14. *Comparison of pitchers of* S. purpurea *ssp.* venosa *on the left, and ssp.* purpurea *on the right. Note that the latter is generally longer and narrower and has a smaller mouth and less expanded hood wings.*

Fig. 3–15. S. purpurea *ssp.* purpurea *f.* heterophylla *growing in an open Michigan bog. Note complete lack of red pigment.*

3–14 3–15

Jersey area, each species of mosquito is able to select and stay with its appropriate plant. However, these initial reports have recently been questioned.

GENERAL.—Looking at the pitcher of *S. purpurea*, one would guess that, of all the pitcher plants, this one would be the least efficient. It apparently has the weakest enzyme secretions and depends heavily on bacterial action for the digestion of its prey, or so present evidence seems to indicate. It traps its prey by drowning it. Since the mouth of the pitcher is widely exposed to weather and flood, rainwater can easily dilute or overflow the contents of the pitcher. But these are relatively synthetic observations which are probably not pertinent in the end. When one sees large populations in good strong bogs and notes the wide area of distribution, one cannot help agreeing that in spite of its clumsy appearance the species has certainly adapted and flourished. We have to learn more about its adaptations.

The most luxuriant stands of *S. purpurea* are in the northern reaches, where there are frequently massive clusters of plants with multiple crowns a meter or more in diameter. The species is also seen growing in dense, floating mats on water at the edges of bog ponds and lakes and nearly all the way across slower, acid streams.

In the southeast, there is an incompletely studied race of the species with large, diffusely red to purple pitchers in sun or even in shade. The flower petals are a pale pink and tend to be sharply curved around the outside of the pale green umbrella, rather than pendulous.

The species is often found adapted to alkaline marl bogs around the Great Lakes, where the pitchers are more numerous, smaller, more brightly colored, and brittle.

Sarracenia flava L.

BOTANICAL NAME: *Sarracenia flava* L.
COMMON NAMES: Yellow trumpet, trumpet, huntsman's horn, lily.
RANGE: The species is generally confined to the southeastern United States in an arc of the coastal plain from Virginia through the Florida panhandle into the Mobile Bay area, but it can be found in some relic bogs in the southeastern piedmont. It has been planted in, and has adapted to, outdoor bogs in Pennsylvania. There are claims of adaptation even farther north, but the year-to-year persistence and quality of the plantings is disputed.

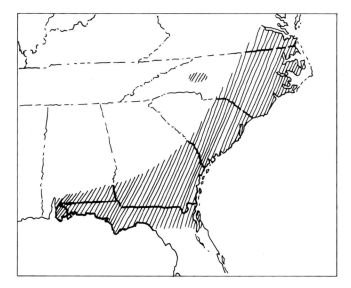

FLOWERING SEASON: Mid-March in the south to mid-April and May in the Carolinas.
TRAP SEASON: The pitchers brown and wither at frost. Straight phyllodia are produced in late summer and persist all winter.

DESCRIPTION.—*S. flava* has an erect, handsome pitcher which grows up to 90 cm. The pitcher has a wide, flaring mouth and a broad, nearly flat, well-formed lid with a prominent keel and a strong column with backward-reflexed margins. The ala is much reduced. This species has considerable polymorphism (vein and color variation), particularly in the Carolinas, with less elsewhere in its range.

There are four typical forms. The most common is pale green to bright yellow when growing in full sun, and there is a large maroon splotch on the inside of the column, from which red veins radiate locally. The next most common form has a bright to deep red color on the external surface of the lid and column, sometimes extending down the pitcher to the ground. Venation is moderate, and the maroon color spot is weak. A third variant is uniformly golden-yellow in full sunlight, with such coarse and prominent red venation all over that the surface has a pleated or reticulate appearance. Again, the interior column spot is weak. Finally, there are plants with no red pigment at all, the mature pitchers being uniformly pale green to yellow. All these basic forms can occur in the same stand, although one or the other frequently predominates, and hybrid intermediates are easily seen. These variants have not been clearly named and are under further study.

The flower is large, and the petals are strap-shaped and bright yellow. There is a very strong feline odor which is noticeable from some distance. The plant produces ensiform (straight) phyllodia in late summer, and these persist through the winter. This fact has not been generally noted in the literature.

GENERAL.—There are areas in Georgia where one can still see magnificent stands of the species—tall, bold, bright yellow pitchers filling a large savannah and melting to a vast golden blur when viewed from a distance. *S. flava* is especially susceptible to the predations of the many insect larvae previously mentioned, but deleterious effects on the species as a whole are not perceptible. Unfortunately, the species does not respond to attacks on its habitat with equal fortitude, and prime lots of *S. flava* are rapidly disappearing, especially in the Carolinas, which were once a main stronghold. This point was illustrated not long ago when an experienced field botanist and I were in what is left of the Green Swamp of eastern North Carolina. We were to finish our walks that day by visiting a favorite location of his for *Sarracenia*. After parking the car and hopping the inevitable drainage ditch, we scoured the savannah for half an hour or so and did not see one pitcher plant of any species. He was standing with a look of bewilderment on his face, and when I came over to him he shook his head and commented, "Where've they all gone? You could always come here when you wanted to see Sarracenias . . . ."

Fig. 3–16. *Yellow, heavily veined form of* S. flava. *Note that the purple area of the throat is not diffuse but is a confluence of veins.*

Fig. 3–17. *Form of* S. flava *lacking any red pigment.*

Fig. 3–18. *Intact flower of* S. flava.

Fig. 3–19. *Typical form of* S. flava. *Note the purple pigment in the "throat" of the pitcher.*

Fig. 3–20. *Red-topped form of* S. flava *beside typical plants.*

33 / The Eastern North American Pitcher Plants

Sarracenia alata Wood

BOTANICAL NAME: *Sarracenia alata* Wood. Unacceptable synonym: *Sarracenia sledgei*.
COMMON NAMES: Pale pitcher plant, flycatcher.
RANGE: This species begins on the Gulf coastal plain, where *S. flava* leaves off in southern Alabama, and continues into east Texas. There is a narrow area of overlap with *S. flava* just east of Mobile Bay.

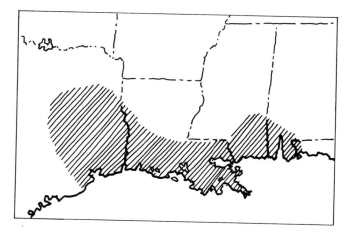

FLOWERING SEASON: Early March through April.
TRAP SEASON: Because of its extreme southern location, the pitchers of *S. alata* often remain over winter in protected areas, withering back with frost in more open places. Year-old pitchers may assume a deep red coloration.

DESCRIPTION.—The pitchers are erect and measure to 75 cm tall. At first glance, this species can easily be mistaken by the inexperienced for *S. flava*, especially in the Mobile Bay area. However, the pitcher mouth of *S. alata* does not flare as widely, the column is not as tall or reflexed, the lid is smaller and more convex, and there is a larger ala (hence the specific name, *alata*).

In good sunlight, the pitcher has a pale yellow-green color, is finely red-veined, and the inner lid and column frequently have diffuse, deep red coloring, as opposed to the purple splotch in *S. flava*.

The flower petals are creamy to yellow-white and are obovate, rather than strap-shaped as in *S. flava*. There is a musty odor much like that of *S. flava*, but it is only about half as strong. There commonly are no winter phyllodia.

GENERAL.—*S. alata* seems quite adaptable and is often found growing in rather dense clay soil as well as in sandy savannahs. It can be seen in large, striking stands on slight slopes and in fields along the highways of southern Mississippi. Most of these western plants have the deep red color in the upper, inner pitchers, and many locations appear literally red with them.

Fig. 3–21. *A clump of* Sarracenia alata.

Fig. 3–22. S. alata.

Fig. 3–23. S. alata *with diffuse, deep red pigment of inner lid and column.*

Sarracenia oreophila (Kearney) Wherry

BOTANICAL NAME: *Sarracenia oreophila* (Kearney) Wherry.

COMMON NAMES: Flycatcher, green pitcher plant.

RANGE: The species is narrowly confined to a few scattered locations in northeastern Alabama. Disjunct sites in middle western Georgia were reported but are no longer extant.

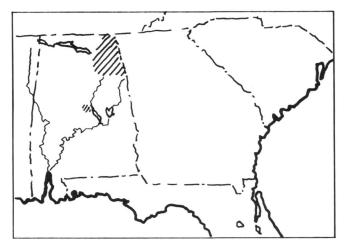

FLOWERING SEASON: Mid-April to early June.

TRAP SEASON: The pitchers wither quite early, usually in midsummer. Curved winter phyllodia are regularly produced in late summer.

DESCRIPTION.—The pitcher is erect and measures up to 75 cm, with a widely flaring mouth, a well-formed column which is not as strongly reflexed as in *S. flava,* and a large lid with a prominent keel. The lid tends to slope slightly more upward than in *S. flava.* The pitchers are very often pale green, but there are stands where there is a fine, red venation. The ala is not prominent.

The flowers have pale yellow-green petals and a very weak odor easily detectable (by nonsmokers), which is best described as mixed sweet and musty (as in *S. purpurea*), the mustiness tending to predominate.

There are phyllodia which appear in midsummer. These are sharply curved and unique for the species.

GENERAL.—This species was thought to be a form of *S. flava* until 1933, when E. T. Wherry discerned that it was not and formally described it. Its range is geographically quite separate from that of *S. flava,* and in fact, no other pitcher plants grow with it except in

Fig. 3–24. Sarracenia oreophila. *Note the fine venation and rather close resemblance to forms of* S. flava, *except for smaller lid and less well-developed column.*

one very small station near Birmingham, where it meets the range of *S. rubra. S. oreophila*'s montane location disjunct from that of *S. flava*, and its consistent morphologic and biologic differences, including the sharply curved phyllodia and the differences in the color and odor of the flowers, all suggest a separate species.

S. oreophila has some interesting adaptive biological features. It grows in wet depressions along streams of the elevated Sand Mountain plateau of northeastern Alabama, where the soil is sandy clay. In midsummer, so-called dog days befall the area, bringing very hot, humid weather with decreased rainfall and considerable drying. At this point, the pitchers begin to brown and wither while other species of *Sarracenia* in more favorable downland areas are still thriving. The prominently curved phyllodia appear and remain green all autumn and winter. The plants catch what prey they can early, then literally fold up for the hot, dry weather. This early withering of the pitchers is largely carried over in cultivation in spite of adequate water and favorable temperatures.

One is not likely to come upon this species very casually, since there were always relatively few stands in a very small range, and these are disappearing rapidly as land is claimed for agriculture and the growth of timber. *S. oreophila* is on the list of very much endangered species and will soon disappear from natural sites. Fortunately, it is adapting in several private and botanical gardens in different parts of the world, in collections made from seeds and plants gathered from condemned or fading areas. But a good natural stand of any species of plant is preferred to a cultivated one. For these reasons, we were especially distressed to learn that a commercial nursery had possibly made a massive illegal collection from an Alabama state park!

Sarracenia minor Walt.

BOTANICAL NAME: *Sarracenia minor* Walt.
COMMON NAME: Hooded pitcher plant.
RANGE: The southeastern coastal plain from the southernmost tip of eastern North Carolina to the mid-Florida panhandle. This is the only species extending into the Florida peninsula.

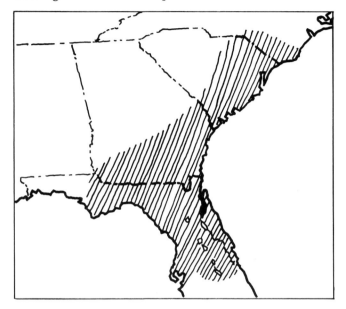

FLOWERING SEASON: Late March to mid-May. This is also the only species in which flowering most commonly occurs simultaneously with or slightly after pitcher growth.
TRAP SEASON: Pitchers tend to persist all winter in sheltered stands in the southernmost range, but elsewhere, they generally die back during severe winters.

DESCRIPTION.—*S. minor* has an erect pitcher averaging about 25–30 cm but growing up to 80 cm in certain areas. There is a prominent ala, and the hood is extremely reflected over and approximated to the elliptical mouth of the pitcher. The column is barely discernible, being quite short, nonreflexed, and continuous with the hood and body of the pitcher. The color is generally green with a coppery red cast to the upper pitcher when it is growing in bright sun. There is fine, red venation over the interior of the hood and the column. Older pitchers frequently turn dark red in late summer. The flower is odorless and has pale yellow-green petals. There are no phyllodia.

GENERAL.—A noteworthy characteristic of this species (and of two others, *S. psittacina* and *S. leucophylla*, yet to be discussed) is the presence, on the back of the upper one-third of the pitcher, of irregular, reticulate, clear to whitish areas that lack chlorophyl. These are called light windows, fenestrations, or areolae. All sorts of functions for the fenestrations in *S. minor* have been suggested, but the most reasonable is that they admit light into the interior of a pitcher rather darkened by the close approximation of the well-developed hood. Insect prey are less likely to fly or crawl into a darkened area than into a lighted one. When alighting or crawling to the top of the broad ala, they may mistakenly confuse the light windows for a place of exit and take off in that direction, striking the pitcher's back wall and falling in. The closely approximated hood, while darkening the interior, undoubtedly provides excellent cover against rain entering the pitcher.

S. minor seems to have a particular affinity for ants. On warm, clear days, one can go to the field and see single columns of ants in two-way traffic up and down the ala, which is studded on its narrow edge with glistening nectar glands. The ants travel all the way to the lip, where many fall in. It is not known whether this represents a specific attraction or whether it is a result of *S. minor*'s tendency to grow in a drier environment where ants are more likely to be.

S. minor is frequently found in the more drained, upland parts of savannahs or in light pine woods, whereas most other species of *Sarracenia* are most abundant in moister, open areas. There is a question as to whether this is a "preferential" adaptation to allay competition with other pitcher plants which do poorly in drier or shadier locations, or whether *S. minor*'s adaptation to a dry, shady habitat simply coincides with the maladaptation of the other species. I have observed and grown *S. minor* for a number of years in both wet and dry soils and in sphagnum, and the plants are much larger and more robust when grown in wetter substrates. In fact, the plants reach their zenith in size in the very wet habitat of the Okefenokee Swamp, where they grow intermixed with *S. psittacina* on huge floating sphagnum islands called prairies. These Okefenokee plants are a special case, and test plants have not been removed and studied in controlled transplant experiments to see if their marked difference in size is really environmental or genetic. I can speak with more assurance with respect to the savannah plants of the rest of the coastal plain, with which I have done transplant studies, and conclude that these plants are hardier in wet substrates, especially in sphagnum moss tubs and plantings, and that therefore the drier location in which *S. minor* is often found is probably not a preferential adaptation.

Although its range overlaps that of many other species of *Sarracenia*, there are places, particularly in South Carolina and parts of the Florida peninsula, where fine, almost pure stands of *S. minor* can yet be seen.

Fig. 3–25. Sarracenia minor, *with flowers in early spring.*

Fig. 3–26. S. minor. *Note the well-developed hood closely approximated to the mouth, and the light windows.*

39 / *The Eastern North American Pitcher Plants*

Sarracenia psittacina Mich.

BOTANICAL NAME: *Sarracenia psittacina* Mich.
COMMON NAME: Parrot pitcher plant.
RANGE: An arc of the southeastern coastal plain from Georgia through the western Florida panhandle into southern Mississippi.

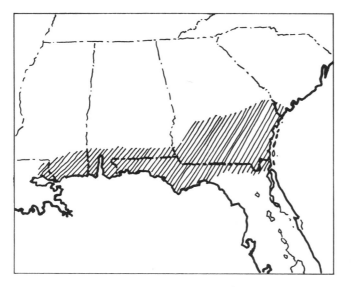

FLOWERING SEASON: Late March to May.
TRAP SEASON: The pitchers remain all year.

DESCRIPTION.—The pitchers are decumbent (except in extreme shading) and grow to 20 cm, although specimens with 30 cm pitchers can occasionally be found. There is a very prominent, undulate ala. The hood is the most elaborate in any species of *Sarracenia*, with the edges united so that it is a globose structure with a relatively small aperture at the top of the ala. Inside, the aperture is surrounded by a 0.5 cm collar, which enables the trap to work somewhat like a lobster pot. The downward-pointing, intermeshing, stiff hairs that line the interior of the pitcher are the most developed in any species of *Sarracenia*. Externally, the coloration is green in the shade to deep red in good sunlight, and light windows are prominent. (See the preceding section on *S. minor* for a discussion of light windows and their possible function.)

The flower is rather small, with petals deep to bright red, and there is a weak but definitely pleasant sweet odor. There are no phyllodia.

GENERAL.—*S. psittacina* has a very distinctive pitcher, its external appearance being rather unlike that of any other member of the genus except in individual morphologic points. A side view of the pitcher discloses a case for the common name, parrot pitcher plant. I have seen the largest, most globose hoods on plants that have been moved to a piedmont Carolina bog, even though the natural range is far to the south.

Fig. 3–27. Sarracenia psittacina. *Note globose hoods.*

This species prefers a very wet habitat, and the prostrate rosettes are often flooded in the spring. This seems to inconvenience the plant little since many aquatic animals are found trapped inside the pitchers. Some botanists feel that flooding is a definite advantage, if not a requirement, for the species. In this situation, one can speculate how the especially large intermeshing interior hairs and the collar surrounding the interior of the aperture are aids in preventing swimming animals from retreating.

As in the case of *S. minor*, this species grows to rather large size on the floating sphagnum prairies of the Okefenokee Swamp, where the two species occur sympatrically. Very large specimens have also recently been found in bogs in southern Mississippi. Again, the question of whether the size factor is environmental or genetic will have to be worked out.

The plant is rather abundant within its range, but it is easily overlooked since it is low on the ground and, by summer, often deep in grass and other plants. The taller flower scapes are a definite help in locating it.

Fig. 3–28. S. psittacina. *Less intensely colored, shade-grown plants.*

Fig. 3–29. *Longitudinal section of pitcher of* S. psittacina. *Note the especially well-developed intermeshing trap hairs inside.*

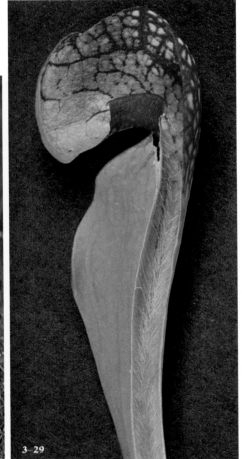

Sarracenia leucophylla Raf.

BOTANICAL NAME: *Sarracenia leucophylla* Raf.
Unacceptable synonym: *Sarracenia drummondii.*
COMMON NAME: White-topped pitcher plant.
RANGE: The southeastern coastal plain from southwestern Georgia into the western Florida panhandle and from the Mobile Bay area to just over the Mississippi line.

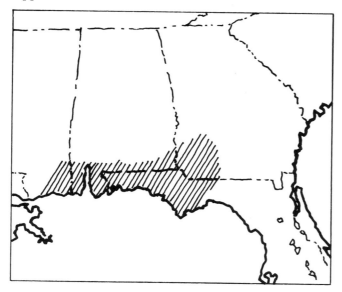

FLOWERING SEASON: Early March to late April.
TRAP SEASON: The traps tend to wither in winter, with new traps appearing in two crops, one concomitant with spring flowering and another, stronger set in late summer. Phyllodia-like structures are often produced during dry periods in midsummer, but they rarely persist over winter.

DESCRIPTION.—*S. leucophylla* has a tall, erect pitcher which grows to 95 cm. The ala is not very promi-

nent except in pitchers that appear very early in the spring and in shade-grown specimens. The most striking characteristic is the white coloration of the hood, column, and upper regions of the pitcher; hence the common as well as the specific name, *leucophylla*, meaning white leaf. There are variations ranging from upper pitchers that are almost pure white with very dis-

Fig. 3–30. *A clump of* S. leucophylla.

crete pale green veins dividing the white area into a mosaic pattern, to pitchers with rather heavy red veins and much red pigment suffused with purple around the white patches. The pitcher mouth is quite expanded, although less so in the variants of more reddish color. The column is well formed and moderately reflexed, and the lid is large and wide, with a prominently undulate margin and large hairs on the lower surface.

The large flower has deep red petals and a sweet odor.

Fig. 3–31. *Two color forms of* S. leucophylla: *white-topped with large mouth and green veins, and a smaller form with red veins.*

Fig. 3–32. *A savannah with moderate growth of* Sarracenia leucophylla, *its white tops easily visible above the grass.*

GENERAL.—There is no doubt that, with the possible exception of bright yellow *S. flava* in now rare massive stands, *S. leucophylla* is the most eye-catching of any species of *Sarracenia*. Its range is rather small, but the plants are still abundant and very frequently massed in large bogs easily observed from many roads. Where I grow a number of species of *Sarracenia* together outdoors in the piedmont area of North Carolina, I have found that *S. leucophylla* is by far the most attractive to insects. After being open only a few days, the pitchers become nearly full to the top. The same observation applies to field specimens of *S. leucophylla*, which generally contain more accumulated insects than do the pitchers of other species. The curious pattern of two crops of pitchers a year is also distinctive.

43 / *The Eastern North American Pitcher Plants*

Sarracenia rubra Walt.

BOTANICAL NAME: *Sarracenia rubra* Walt. Also, a generally acceptable subspecies: *Sarracenia rubra* ssp. *jonesii* Wherry.

COMMON NAMES: Sweet pitcher plant, red pitcher plant.

RANGE: In an arc throughout the coastal plain of the southeastern United States, except the Florida peninsula, and into southwestern Alabama. The disjunct subspecies is in the southwestern mountain counties of North Carolina and one adjacent county in South Carolina.

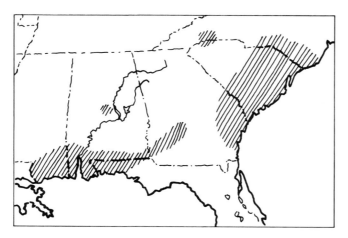

FLOWERING SEASON: April to May in the coastal plain, to June in the Carolina mountains.

TRAP SEASON: The pitchers wither with frost.

DESCRIPTION.—This is a difficult species to discuss. There is much animated controversy over the exact taxonomic status of the populations, which some botanists would classify as a complex of species. We will try to present a practical summary of the situation.

The pitchers are erect and grow up to 50 cm tall when fully developed in ideal habitats. Mature pitchers generally have bright red, fine veins on a coppery tan background; in less mature specimens the red tends to be more diffuse. The mature summer pitchers have a moderately prominent ala. In the typical coastal plain form, the column is short and nonreflexed while the lid is elongate, well developed, and somewhat closely approximated to the elliptical pitcher mouth, although not nearly as closely as in *S. minor*.

In the Carolina mountains, where the controversial *jonesii* subspecies is found, the pitchers grow 15–20 cm taller than in the coastal plain form. They have better-developed, wider hoods, taller columns with some reflexion, and a distinct bulbous widening of the upper quarter of the pitchers, which can be seen when the pitchers are viewed from the side.

In the western Gulf coastal plain area, there is a variant form, which is as tall as the mountain specimens but is a more diffuse red in sunlight and has a hood which is better developed, more elongate and convex, and often somewhat undulate at the margins. The column is slightly reflexed and better developed than in the eastern plants, but not as strongly developed as in the mountain *jonesii*.

In nearly disjunct locations around Montgomery, Alabama, and extending down to northern Baldwin County, there is an additional variation, measuring to 30 cm, with stockier pitchers and wider openings. There is only a faint red coloration of the lid and very fine venation of the upper pitcher when the plant is grown in full sunlight.

Rather regularly, plants of this entire complex species have a two-stage leaf sequence. Spring leaves may be somewhat phyllodiform, with a broad ala and a small pitcher tube. The leaf is often deformed, being

Fig. 3–33. Sarracenia rubra (*typical form of eastern coastal plain*) *in flower in early spring. Note the dead pitchers from previous years at the left, and younger pitchers emerging.*

Fig. 3–34. S. rubra, *same pond edge as above, later in season. Petals have fallen from the flowers, leaving the umbrella and sepals. The pitchers are now better developed and well colored.*

Fig. 3–35. *A form of* S. rubra *peculiar to the Gulf coastal plain.*

Fig. 3–36. S. rubra *ssp.* jonesii. *Note the more robust pitcher tops and the definite bulge or widening just below the mouth.*

ensiform, or roughly S-shaped. The summer leaves, which appear later, are more typical of the particular form. This dimorphic pattern varies considerably among different populations and forms because of genetic differences, local growing conditions, or a combination of these factors. Reciprocal transplanting of individuals may elucidate the mechanism involved.

The flowers are rather small and have bright red petals. While most species of *Sarracenia* have only one flower per growth crown, *S. rubra* frequently has multiple flowers in all its forms. The pleasantly sweet odor is strong except in the Gulf coastal variants, which have only a weak odor. There are no true winter phyllodia.

GENERAL.—Part of the problem with the taxonomy of this attractive little plant could be put into perspective if more consideration were given to the current tendency to consider species as dynamic plant populations. Plants are always interacting with each other and with their environment and are fully involved in evolution and its changes; they are not the static entities that symbolic names suggest. An excessive desire for certainty and a need to label can lead to quibbling which suppresses truly important questions. In most cases, plants do not change or evolve much in one man's lifetime, or even in many lifetimes, and it is difficult to appreciate broader concepts than those which are immediately apparent.

S. rubra appears to be exhibiting something of a sprawl of evolutionary divergence and probably illustrates the concept of incomplete differentiation. Perhaps the seemingly equivocal but really quite useful terms *semispecies* and *syngameon* should be considered by more students of extreme persuasions as a replacement for the term *subspecies*. (Curiously, the variations of *S. flava*, and even the lesser ones of *S. oreophila*

and *S. leucophylla*, are just as striking and probably as important as those in *S. rubra*, yet they have received little or no attention.)

In 1929, E. T. Wherry inadvertently sparked this controversy when he published his description of the new species *Sarracenia jonesii* for disjunct populations of *S. rubra* in the Carolina mountains. He felt that geographical separation from the rest of the *rubra* stands, along with the morphological differences and his incorrect observation that the flower of the mountain population had no odor, were all sufficient to suggest a new species, somewhat parallel to his separation of *S. oreophila* a few years later. Unfortunately, he apparently made the mistake of believing that the more robust Gulf coastal variants of *rubra* and the disjunct central Alabama populations* were identical with the mountain plants, which they are not. He has since rescinded his opinion, but *S. jonesii* is recorded in botanical literature as being in the Gulf coastal plain. The *S. jonesii* designation of the Carolina mountain plant met with considerable resistance in the forties, and recent monographs on *Sarracenia* have since tended to relegate the mountain plants to either subspecific status or no status at all.

The confusion was further compounded when the typical *S. rubra* of the eastern coastal plain was reported in mountain bogs alongside *jonesii*. I am satisfied that there are no coastal plain forms of *S. rubra* in these mountain slope bogs, and that there never were. The immature pitchers of younger or disturbed plants of *jonesii* look like the coastal ones until the

*A recent paper (Case, F. W. and Case, R. B. 1974. *Sarracenia alabamensis*, a newly recognized species from central Alabama. *Rhodora* 76: 650.) suggests that the separate central Alabama plants—also becoming extinct—should be designated as a species, *Sarracenia alabamensis*. The authors' studies were not thorough, however, and my impression is that these populations will likely prove to be an *S. rubra* subspecies.

plants mature or acclimate. The report of both types and their intermediates was one of those misobservations that had not been followed up by the necessary years of carefully observing marked field plants and reciprocal transplants.

Still more recently, two types of eastern coastal plain forms of *S. rubra* have been suggested! Again, this is a hypothesis based on static herbarium studies and irregular field visits, rather than on long-term observations of live, growing plants. The pitchers simply grow taller and are more robust in plants growing in wet sphagnum areas, on pond edges, and in low sandhill bogs than they do in the smaller, almost miniature forms growing in drier savannahs. This has been proved to my satisfaction by reciprocal transplants as well as by growing the plants in outdoor tubs and a greenhouse for several years. The differences reported in eastern coastal plain plants are, therefore, nongenetic.

For the time being, I would suggest the following guideposts for the beginner and even for more advanced naturalists who are equally confused by the jigsaw picture in botanical literature. Taking *S. rubra* in perspective with all other morphological and biological considerations in the genus and with relative differences between accepted species of *Sarracenia*, the basic designation *S. rubra* can apply to all populations, with a subspecific designation *S. rubra* ssp. *jonesii* for the Carolina mountain plants, since consistent morphological and biological differences along with the factor of isolation cannot be ignored. But none of these are sufficient in themselves or together to suggest a separate species for *jonesii*. The Gulf coastal variants and the central Alabama plants also deserve a subspecific status, although some botanists are willing to concede species status, which I believe would be premature at this point.

The whole problem may, before long, become an exercise in futility. The *jonesii* plants are nearly all extinct. They can be found in only a few small bogs after considerable search where they were once easily seen in masses from the roadside. Likewise, the typical coastal plain plants are fast disappearing in the east, particularly in the sandhills below the piedmont.

SARRACENIA HYBRIDS

The maps showing the ranges of species of *Sarracenia* indicate that many occur in the same or overlapping ranges. In the field, one will often note that two or more species occur in the same bog or savannah. As a result, crosses or hybrids between two species are frequent.

Not all species in the plant kingdom are capable of hybridizing. When species do cross, the resulting progeny are frequently sterile—that is, they are incapable of further sexual reproduction. Such is not the case with *Sarracenia*, however. Not only can nearly all conceivable crosses be found or made in the field and greenhouse, but the hybrids are quite fertile and are capable of forming additional crosses with third species or with each other, or they can effect complex backcrosses with one or both parent plants. Such backcrossing (introgressive hybridization) results in an exchange of genetic material between more or less established species, and it is felt by many botanists that this is an important factor in creating variation for evolution.

There is a general rule in botany that plant hybrids become established only with difficulty, since in theoretical ecologic terms (and quite usually in actuality) they should require a rather narrow habitat intermediate between those of the two parents, which have themselves become established as a result of specific environmental selection. But the various species of

3-37

3-38

3-39

3-40

3-41

Sarracenia can generally grow pretty well in similar environments, so the hybrids establish themselves intermingled among the parents, provided there are physical space and proper conditions for seed germination along with a minimum of competition from other species in the bog. The hybrids may, however, come to occupy newly opened, disturbed areas.

One might ask, then, if these species are so capable of crossing with each other, how distinct species have been preserved so well, especially in sympatric bogs. The answer lies partially in isolation factors. Isolation most commonly involves physical or geographic barriers or displacements. There can also be reproductive isolation, such as a genus' inability to hybridize, or the formation of sterile hybrids or hybrids that may be ecologically incompatible with the locality where their seed is shed. But we have seen that neither of these factors quite applies to species of *Sarracenia* occurring in the same bog, although the physical separation factor would apply in disjunct bogs of single species and at the far ends of ranges.

When two or more species of *Sarracenia* occur intermingled in the same bog, other somewhat more subtle isolation factors come into play. The main one appears to be differences between the peak flowering periods of the species, which means that when the pollen of one species is ripe, the stigma of another species may not be receptive yet, or the plant may not even have flowered. Although flowering periods sometimes overlap, there is relatively little opportunity for crossing, since the two *peaks* of flowering activity within the broad flowering periods are different in most cases. Thus, there would be few plants of sympatric species able to cross at the critical period. Such an incomplete isolation factor is frequently referred to as "leaky."

There are some other equally leaky isolation factors that apply to *Sarracenia*. A large pollinator capable of negotiating a large-flowered species may not be able to enter the flower of a smaller sympatric species, although the reverse could occur. Color and nectar odor may attract certain pollinators preferentially, thus limiting the transfer of pollen between species with different flower colors or odors. Many pollinators exhibit a characteristic known as fidelity, whereby a particular colony of pollinating insects will visit only one species, or even one stand of a species, until it ceases to flower, rather than visiting all flowering plants indiscriminately.

Plant dispersal in *Sarracenia* in general is also limited. Although much more field research needs to be done on this subject, the *Sarracenia* pollinators that have been studied are not far ranging. Furthermore, the seeds of the genus depend on floods or nearby flowing water to be dispersed. They are too heavy for wind dispersal, and we do not know of any birds or mammals that carry them. Other isolation factors common to the entire genus involve the inability of a hybrid seed crop to become established in a crowded area without open, disturbed sites and without suitable soil and water after seed dispersal does occur. Finally, many insect larvae attack maturing seed pods.

Individually, these isolation factors are seen to be fraught with loopholes, but if they are taken together with observations in the field, there is good evidence that they are generally effective, even though many hybrids are rather easily found.

Fig. 3–37. S. flava *x* S. purpurea *hybrid.*

Fig. 3–38. S. purpurea *x* S. rubra.

Fig. 3–39. S. minor *x* S. purpurea.

Fig. 3–40. S. leucophylla *x* S. purpurea *x* S. leucophylla.

Fig. 3–41. S. leucophylla *x* S. psittacina.

While inspecting healthy locations of *Sarracenia,* one will notice that certain hybrid combinations occur with variable frequency depending on the application of any or all of the possible isolation factors mentioned above, and probably on many other factors we do not yet know about. The hybrid between *S. flava* and *S. purpurea* may be found with moderate ease in a location where the two species occur together. The hybrid between *S. flava* and *S. minor,* on the other hand, is relatively rare where the two species grow together. Some species hybridize so freely where they occur together that so-called hybrid swarms are found. This is frequently the case with *S. alata* and *S. leucophylla,* where several large bogs full of intercrosses and backcrosses yield few plants that can be identified as pure parents. These bogs occur at the narrow interface of the ranges of the two species, and it is clear that geography plays a big role in keeping these two species intact.

The frequency of hybridization in *Sarracenia* caused a great deal of difficulty among early botanists, who often thought that each hybrid was a new species and so named it. Over years of study, the species were sorted from the hybrids, but some of the original species names given to the hybrids stuck in a modified way, and the plants are often referred to accordingly. For example, a plant named *Sarracenia catesbaei* was later found to be a cross between *S. flava* and *S. purpurea.* The hybrid should most properly be written *S. flava* x *S. purpurea,* the x reflecting the hybrid status and pronounced "by" (colloquially "times") when spoken. This system, although accurate and to the point, is clumsy in rapid conversation or reading, especially when one gets into hybrids composed of three or four or more ancestors! So, the old mistaken species names have been informally allowed (and some new ones have been added in modern times, largely in

vanity) as *horticultural names,* with the placement of an *x* in the written version, as in *S.* x *catesbaei,* the *x* warning that we are not dealing with a species.

In appearance, *Sarracenia* hybrids assume a form that is generally intermediate between those of the two parents, rather than representing a patterned mosaic or dominant-recessive situation. A yellow-flowered species crossing with a red-flowered species does not produce a hybrid with one color or the other, or with spotted flowers; it produces pink or orange flowers. A species with erect pitchers crossing with a decumbent species produces a hybrid with pitchers that are semi-erect.

There are large numbers of natural hybrids that have been found and recorded, and a few possible ones are yet to be found, but practically all possible combinations have been produced in the laboratory or greenhouse. We have included photographs of only a few of the naturally found hybrids as examples. The interested reader is invited to consult some of the references for more discussion.

Some naturally found, simple, two-parent *Sarracenia* hybrids and their horticultural names:

Botanical name	Horticultural name
S. flava x *S. purpurea*	*S.* x *catesbaei*
S. leucophylla x *S. purpurea*	*S.* x *mitchelliana*
S. minor x *S. purpurea*	*S.* x *swaniana*
S. psittacina x *S. purpurea*	*S.* x *courtii*
S. purpurea x *S. rubra*	*S.* x *chelsoni*
S. alata x *S. purpurea*	*S.* x *exornata*
S. flava x *S. leucophylla*	*S.* x *mooreana*
S. leucophylla x *S. minor*	*S.* x *excellens*
S. leucophylla x *S. psittacina*	*S.* x *wrigleyana*
S. leucophylla x *S. rubra*	*S.* x *readii*
S. alata x *S. leucophylla*	*S.* x *areolata*
S. flava x *S. minor*	*S.* x *harperi*

S. minor x S. psittacina   S. x formosa
S. minor x S. rubra   S. x rehderi
S. flava x S. rubra   S. x popei
S. psittacina x S. rubra   S. x gilpini
S. alata x S. rubra   S. x ahlesii

DIFFICULT IDENTIFICATIONS

With some experience and care, one will have very little trouble identifying species of *Sarracenia* in the field. The only three species that closely resemble one another in some ways are *S. flava*, *S. oreophila*, and *S. alata*. *S. oreophila* is completely restricted to its range in northeastern Alabama and is becoming so rare that the casual observer is unlikely to come upon it except in collections of live plants. The pale flower and sharply curved phyllodia are characteristic. *S. flava* and *S. alata* can be found in the same bogs in a few instances around Mobile Bay, where their otherwise separate ranges intersect. A careful comparison of the photographs will show that *S. flava* has a much more pronounced and reflexed column, a larger and flatter lid, and a more widely flaring mouth. Flower differences are also present and can be used for identification if one comes upon the plants in that stage: the petals of *S. flava* are bright yellow and strap-shaped, and there is a strong feline odor, while the petals of *S. alata* are a paler yellow and more rounded, and the musty odor is far less strong. When *S. alata* is growing in good sunlight, the external surface of the pitcher is more likely to be a diffuse, pale, yellow-green with fine red veins, and in many examples of the plant the uniformly dark red color of the inner lid and column is distinct from the purple splotch and coarser vein patterns of typical examples of *S. flava*.

In larger bogs, especially on the Gulf coastal plain, one will surely come across hybrids. Some of the more complex combinations confound even experts who attempt to analyze parentage by simple inspection. Generally, the best approach is to note carefully what species of potential parents are nearby. (This will often not work very well along drainage ditches!) Then, remembering the tendency of *Sarracenia* hybrids to have an appearance intermediate between those of the two parents, one should try to pick out the species characteristics in the hybrid. This can be quite interesting, and with some experience, you will gain perspective enough to analyze readily most hybrids.

IV. The California Pitcher Plant (*Darlingtonia californica* Torr.)

BOTANICAL NAME: *Darlingtonia californica* Torr. Unacceptable synonym: *Chrysamphora californica*. Family Sarraceniaceae.
COMMON NAMES: California pitcher plant, cobra plant, cobra lily.
RANGE: Pacific coastal bogs and mountain slopes from Oregon to northern California. Altitude varies from sea level to 2800 m.

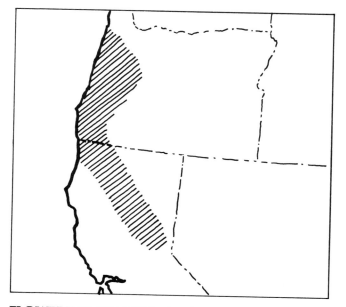

FLOWERING SEASON: April to August, depending on altitude.
TRAP SEASON: Traps tend to remain over winter if moderately protected.

DESCRIPTION.—*Darlingtonia* has mainly erect to sometimes decumbent tubular pitcher leaves which grow up to 90 cm, the semidecumbent leaves tending to be smaller. The pitcher leaves are narrow at the bottom and widen to 12–15 cm in a somewhat globose hood at the top. At the top of the moderate ala is the trap opening, which faces downward and is rather large, measuring up to 3 cm. An apronlike, two-lobed, "fishtail" appendage projects downward from the outside edge of the pitcher opening opposite the ala. From a side view, the whole effect is that of a fancied cobra with expanded hood and a rather large, protruding, forked tongue.

Whereas in the genus *Sarracenia* the pitcher openings all tend to face the center of the rosette, the pitchers of *Darlingtonia* twist 180° in either direction as they grow, so that the pitcher openings always face away from the center of the rosette.

The pitchers are mostly pale yellow-green above and a darker green below. In full sunlight, there is frequently much red to yellow coloration of the upper portions of the pitchers. The hood area has numerous confluent light windows, or fenestrations. (See discussion of fenestrations under *Sarracenia minor*, page 38.) There are nectar glands over much of the upper external surface of the leaf, and these are especially well developed on the tonguelike appendage.

The inner margin of the pitcher opening is rolled into a small collar, and this feature, in addition to numerous downward-pointing hairs lining the hood and depths of the pitcher, discourages the escape of prey.

There is a smooth, hairless zone between the hood and the bottom of the pitcher, however. *Darlingtonia* has no digestive glands. It is presumed that prey is decomposed by microorganisms and that the nutrients are then absorbed directly by the lining cells. There are no intrinsic enzymes in the pitcher fluid, but the presence of prey, together with certain chemical or mechanical stimulations, does increase the secretion of water from the lining cells into the pitcher cavity.

The plant is perennial, with a long, branching rhizome and fibrous roots. Vegetative reproduction is more prevalent than sexual reproduction. The numerous stolons (runners) which grow from the rhizome are probably most responsible for the massive proliferation of *Darlingtonia* at suitable locations.

The flower has a tall scape which assumes the form of a shepherd's hook at the time of anthesis (opening). There are several bracts (leafletlike blades) at intervals on the scape. The bracts are colored from pale yellow-green to pink to deep red. The actinomorphic flower has five elongate green sepals, which project horizontally or reflex at their bases, and five pendulous crimson petals, which come to a point and are closely approximated, the result appearing to be a closed, conical corolla. Near the tip of each petal, the lateral borders have semicircular notches, so that, when the petals are seen together, there appear to be five circular openings into the corolla.

Inside the corolla, the large ovary is bell-shaped, and the very short style and five-lobed stellate stigma project down from the flat, wide bottom surface of the bell. The twelve to fifteen short stamens are arranged around the narrower base of the bell. This anatomical structure serves to encourage cross-pollination. A pollinator may enter the flower through the circular openings in the corolla or may separate the tips of the petals. It will likely brush over the stigma at this time

Fig. 4–1. *A clump of* Darlingtonia. *Note the rotation of the pitchers as they have grown, and the large hood.*

and deposit pollen collected from other flowers. The pollinator then ascends the bell-shaped ovary to the flower base, where it may collect additional pollen. As it leaves the flower, the sloping configuration of the ovary tends to prevent the pollinator from touching the stigma again, this time with the flower's own pollen. Also, the bell shape prevents pollen from falling directly from the stamens onto the stigma, which is largely shielded under the wide bottom surface of the bell.

The seeds set by autumn. They are pale brown and elongate, up to 2 mm in length. The bulbous end of each seed is covered with numerous short projections, which may indicate dispersion by animals.

GENERAL.—*Darlingtonia* is quite distinct from the eastern *Sarracenia*, although it is a member of the same family. Large numbers of these plants growing in great, cool, green, western bogs are equally as attractive as eastern pitcher plants.

The species was discovered in 1841 by J. D. Brackenridge, a botanist assigned to an exploratory expedition in California. The expedition was constantly threatened with hostile attack, but Brackenridge persisted in wandering from the protection of the main force in order to botanize. He came upon *Darlingtonia* and had only a few seconds to tear off portions of a few leaves and a seed capsule or two before he had to rush back to the group. Back in the east, the renowned botanist John Torrey immediately realized that the species was new and that it was related to, but different from, *Sarracenia*. He formally described and named the plant from the scant material hurriedly assembled.

Several years later, a specific search was made for the plant, and it was found in several areas. It was most intensively and lovingly studied by a Mrs. Austin, who grew up in the late nineteenth century in the area of the Feather River in California. As a child, she had guided some of the botanical explorers after Brackenridge to stands of *Darlingtonia* located in an adjacent valley, which she herself had explored. Although not formally trained, Mrs. Austin spent several decades studying *Darlingtonia* and submitting detailed notes to the renowned eastern botanist Asa Gray. These notes remain to this day the most complete and exhaustive field study records available on the species. It was Mrs. Austin who determined the nature of the pollination process and the lack of digestive enzymes, which was later confirmed by the entomologist Frank Morton Jones. She even braved a violent mountain thunderstorm to sit among the plants and observe that the hoods do indeed effectively prevent the entry of rainwater into the pitchers.

Darlingtonia grows in sphagnum bogs or in poor peat soils and gravel near springs and cool, fast-running streams. The geologic base rock is serpentine, which has very poor nutrient value. While summer days may become quite warm, the rhizomes and roots are always immersed in boggy slurries that are constantly permeated with cold spring waters which seldom exceed 20°C. The plants occur only rarely in standing-water bogs and seem to do best in—if not require—cool, moving water at the root level. They may thus be seen growing in seepage bogs and springheads, along streams and ponds, and even with minimal foothold in the snags of rapids and waterfalls.

The chief thing that attracts prey seems to be the gland-bearing "fishtail" hood appendage, which is present in modified form even in the earliest seedlings. The appendage attracts not only flying prey to erect pitchers but ground prey to decumbent pitchers as well, when the pitchers are bent and twisted so that the lobe touches the ground. Once, while doing field studies on the species, Frank Morton Jones was carrying an armload of pitchers back to camp when he noticed a butterfly fluttering near and finally landing on the lobe of one of the pitchers, even as he was walking.

Like *Sarracenia*, *Darlingtonia* has several insect associates, the most common of which is the larva of a gnat, *Metriocnemus edwardsi*, which lives in the pitcher liquid but apparently does no harm to the plant.

Fortunately, the California pitcher plant seems destined to be with us for a long while, since at the moment it is not seriously endangered. Many good *Dar-*

lingtonia areas are parts of state and federal reserves, and the rough, poor land and stunted, deformed trees of mountainous areas discourage agriculture and forestry. The difficulty of access to many of the bogs prevents their being overrun with visitors or easily vandalized. However, even some of our most remote and primitive areas are lately suffering increased use by weekend pioneers, and there is a growing market for the species in the nursery trade.

Fig. 4–2. *An intact flower.*

Fig. 4–3. *A flower with two petals removed. Note the notching of the remaining petals, the large bell-shaped ovary, and the stamens ringing the base of the ovary.*

Fig. 4–4. *Close-up of the upper portion of a pitcher. One can clearly see the fishtail appendage, the pitcher opening, and many light windows.*

V. The Sundews (*Drosera* L.)

The Genus

BOTANICAL NAME: *Drosera* L. Family Droseraceae.
COMMON NAMES: Sundew, catch-fly. (Many species do not have individual common names.)
RANGE: Mountain and coastal bogs of the Pacific northwest into northern California; and, generally, in bog locations throughout the eastern third to half of North America.
FLOWERING SEASON: Spring and summer; varies somewhat with species.
TRAP SEASON: Varies with species; most form winter hibernacula. (See discussion below.)

DESCRIPTION.—These are generally perennial herbaceous plants, although they sometimes have an annual cycle and reappear through seed germination the following season. The plants all form a rosette pattern and have fibrous roots and a stem of variable length which may rise above the ground. The leaves are erect or prostrate, depending on species and situation, and with the exception of one species, the stem is quite short. The leaves are of two parts: a narrow, linear petiole (leaf stem) of variable length, and a terminal blade modified into a trap.

The trapping portion of the leaf varies in shape from circular to linear to filiform (threadlike) in different species. It is always flat and has numerous stalked glands, mostly over the upper surface, with fewer on the lower. The glands are frequently bright red and have a dewdroplike secretion in humid situations. Often, the entire leaf blade seems to be glistening with dew—hence the common name and the generic name, from *droseros*, which, in Greek, means glistening in the sun. The stalked glands are of two main types. Those on the periphery of the leaf have longer stalks and function mainly in entrapment. Those near the centers of broader leaves are shorter-stalked to sessile (stalkless) and function in further entrapment, but mainly in secreting digestive fluids. Leaves are produced continuously all season.

Prey, usually very small insects, is perhaps lured to the trap leaves by the coloration and the sweet nectar secretions of the glands, or the flying insect may simply be seeking a landing platform. When crawling or alighting on a leaf, the prey becomes mired down in the sticky secretions, and the longer-stalked peripheral glands then slowly bend in to the center of the leaf, placing and securing the prey in the digestive area of the sessile glands. The mechanism for bending the gland stalks is not understood, although electrical activity is involved. In some species a very slow leaf folding takes place, tending to secure the trapped prey even further. This leaf folding movement is most clearly observed in *D. rotundifolia* and *D. intermedia*.

The flowers are on a tall scape arising from the stem between two leaves. The form of the inflorescence most closely approximates a raceme, since each flower has a short pedicel between its receptacle and the scape, as does a lily of the valley; but it is commonly referred to as a spike. There are on a spike from five to thirty flowers lined consecutively down one side. The lowest flower opens for one or two days, then closes, and the others follow suit in nearly daily succession upward to the tip of the spike. The actinomorphic flowers are rather ordinary, the parts being mainly in multiples of five: five sepals, five petals, five to ten stamens, and a five-part ovary. There are three two-lobed styles. The flowers average 0.3–1.0 cm across, but the flowers of some species may be as large as 2.5 cm. The petals may be white to rose pink. Petal color is not a reliable way to differentiate species in this genus, with the exception of D. filiformis, which typically has rose-pink petals.

During the day, when the flowers are open, cross-pollination is effected by wind or small insects. If cross-pollination does not take place during the day, self-pollination occurs in most species as the flower closes at night, thus insuring the continued production of seed.

The seeds are black and elliptical and are less than 1 mm long. Their surface patterns are consistent for each species, and in fact, with the aid of a microscope, one can identify the species by its seed.

GENERAL.—Worldwide, there are nearly a hundred species of this genus. Sundews have proved quite durable and seem to withstand the rigors of environmental misuse better than most other carnivorous plants. Indeed, a disturbed site where other native vegetation has been destroyed provides an ideal opening for the colonization of seedlings. Drosera is among the first plants to come back in cutover sites and after roadside ditching and burning.

While many of the North American species are rather small, some reach spectacular size. D. filiformis v. tracyi, along the Gulf coast, for instance, has leaves a half meter long. Even the smaller species are fascinating when seen in masses. Early on a bright morning their dewy, sticky tentacles shine and glitter with all the colors of the spectrum. Mats of at least one species grow floating on the edges of, and in some places nearly across, slow, acid streams and the water of drainage ditches in the southeast. In other places, roadside banks are so red with sundews that the plants might be mistaken by a passing motorist for a raw clay surface.

An interesting feature of some species is the autumn production of hibernacula, or winter buds. This occurs in species that grow exclusively in the mountains or in the north, or in species whose ranges are mainly northern but may extend into the southeast. The species of Drosera found only in the southeastern coastal plain do not produce hibernacula. The winter bud is actually a tight, somewhat spherical grouping of leaf primordia (budlike young leaves), often markedly hirsute, from which the plant will grow again in the spring. A hibernaculum is evidently better able to withstand cold weather than is an open rosette, and its formation serves to protect the plant from total destruction over winter. After the formation of hibernacula occurs, the remaining leaves and frequently the roots die back, leaving no trace of the plant unless one has a sharp eye for finding the winter bud among debris.

The Species

Drosera rotundifolia L.

BOTANICAL NAME: *Drosera rotundifolia* L.
COMMON NAME: Roundleaf sundew.
RANGE: In the west, from Alaska south through British Columbia, western Montana, and northern California; in the east, from Labrador and Newfoundland west to the Great Lakes and south into the Appalachians. A reported location for this species in the Carolina coastal plain is not confirmed at this writing.

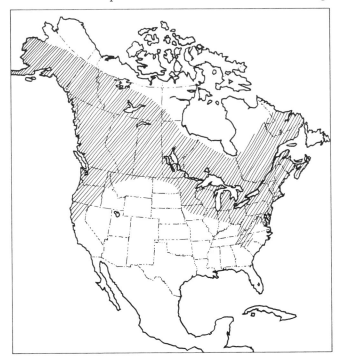

FLOWERING SEASON: June to September.
TRAP SEASON: Spring and summer. Forms hibernacula.

Fig. 5–1. Drosera rotundifolia, *whole plant in flower.*

Fig. 5–2. D. rotundifolia, *close-up of leaf blade. The blade is wider than long.*

DESCRIPTION.—The plant rosette averages up to 8 cm across and becomes ungainly where it grows in deep, coarse sphagnum. The petioles are long, up to 4 cm. The leaf blade is up to 1 cm across and can be round but is usually wider than it is long, an important characteristic for identification. The flowers are more often white than pink.

GENERAL.—Charles Darwin devoted nearly half of his book *Insectivorous Plants* to this species. He performed numerous physiological experiments and observations and recorded his results with his usual attention to intricate detail. This is certainly one of the most widespread species of *Drosera* and occurs in Europe as well as in America.

The plant can usually be found in sphagnum bogs, and often coarse tufts of moss have grown up so that the trap leaves are barely visible. If one stoops down and separates gently the strands of sphagnum, the rest of the plant is often disclosed. The largest and best developed plants of this species that I have ever seen are in the New Jersey Pine Barrens and in some bogs in Oregon, where the leaf blades are dime-sized.

Drosera linearis Goldie

BOTANICAL NAME: *Drosera linearis* Goldie.
RANGE: From Labrador west into the Great Lakes area and south to Michigan.

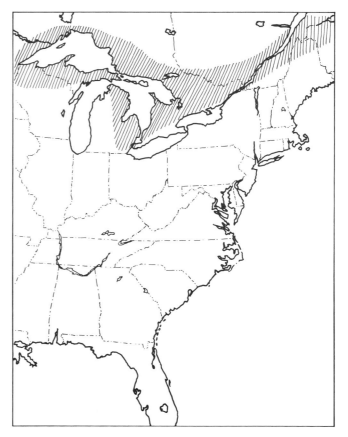

FLOWERING SEASON: June to August.
TRAP SEASON: Spring and summer. Forms hibernacula.

59 / *The Sundews*

DESCRIPTION.—*D. linearis* has generally erect leaves with 2 cm petioles and gland-bearing blades which grow up to 3 cm. The leaves are quite elongate—hence the specific name, *linearis*. The flowers are usually white.

Fig. 5–3. D. linearis, *whole plant.*

GENERAL.—This interesting species is restricted to cool regions of the border area between the United States and Canada. It quite characteristically grows in alkaline marl bogs, rather than in the acid situations that are more common for most carnivorous plants. Very frequently, it is accompanied by *Sarracenia purpurea*, which can grow in either acid or marl conditions in the north, and by two other acid-preferring species

Fig. 5–4. D. linearis, *single leaf. The blade is long with parallel sides.*

of *Drosera*, *D. rotundifolia* and *D. anglica*, which grow on the sides and tops of acid, mossy hummocks or tufts scattered like little islands over and around the edges of the wetter, marly areas. In this situation, both *D. linearis* and the acid-loving pair are able to grow in respectively suitable alkaline or acid habitats in the same bog. Plants of *D. linearis* will even grow in shallow water, if it does not cover the leaves and is not acid.

The matter of the plant's unique occurrence in marl conditions has often been looked into superficially, but the basic question of whether the species' adaptability to alkaline soils is obligatory or a secondary advantage has not been resolved. Several other people and I have found that in culture, the seeds germinate and the plants grow as well in acid, sand-peat soils and in sphagnum as they do in more alkaline conditions, yet in nature, the species is only rarely found ascending the acid, mossy hummocks which are frequently found in marl bogs and which support several species of acid-loving plants. Perhaps *D. linearis* is unable to compete with the inhabitants of the more acid microhabitats and takes advantage of its adaptability to marl conditions where few other herbaceous plants grow and compete for space.

The species is on a marked decline, particularly in recently well-documented locations in southeastern Michigan. The decrease seems to be correlated with the deterioration of marl bogs to acid conditions and with a string of persistent rainy seasons with flooding to the extent that the plants of *D. linearis* were totally submerged. While the species does grow in shallow marl waters to a depth of a centimeter or so, it will not tolerate prolonged flooding. Other possible factors leading to deterioration are toxic pollutants entering natural water systems through runoff or as lands are developed; and, of course, the total drainage of a bog.

Drosera anglica Huds.

BOTANICAL NAME: *Drosera anglica* Huds. Unacceptable synonym: *Drosera longifolia*.
RANGE: In the west, from the Aleutians south to Alberta, western Montana, and into northern California; in the east, from Labrador west into the Great Lakes area.

FLOWERING SEASON: June to August.
TRAP SEASON: Spring and summer. Forms hibernacula.

DESCRIPTION.—The leaves are semidecumbent (partially reclining) with rather long petioles reaching 3 to 4 cm. The blades are pale green with bright red,

stalked glands, and are longer than wide, measuring to 2.5 by 1.0 cm. The flowers are most often white.

GENERAL.—This very attractive sundew is especially interesting to botanists because there is pretty good evidence that it has evolved from two other contemporary species, *D. linearis* and *D. rotundifolia*. The simple hybrid between these two, frequently found where they are sympatric, is sterile. However, if the number of chromosomes of the cells of the hybrid embryo is doubled through an accident of cell division (amphiploidy), then the flowers of the plant growing from this embryo will be fertile, and the plant will reproduce sexually true to species. The sterile hybrid and the fertile hybrid (*D. anglica* species) look very much the same outwardly, although microscopic examination discloses larger cells in the amphiploid specimens. In view of the origin of the plant, some botanists would prefer to write it as a hybrid all the time (*D. x anglica*), rather than as a species, and they would simply note whether the particular plant is sterile or fertile. There remains one very perplexing problem: given the geologic sequence of events in North America, how does *D. anglica* happen to occur in the west and even in Europe, where no *D. linearis* has ever been recorded?

This bright sundew occurs in acid sphagnum bogs, or on acid mossy hummocks in marl bogs, frequently in the company of its probable ancestors.

Fig. 5–5. D. anglica, *whole plant.*

Fig. 5–6. D. anglica. *The leaf blade is somewhat oval, is pale green, and has red glands.*

Drosera intermedia Hayne

BOTANICAL NAME: *Drosera intermedia* Hayne.
RANGE: In suitable locations over most of the eastern third of North America.

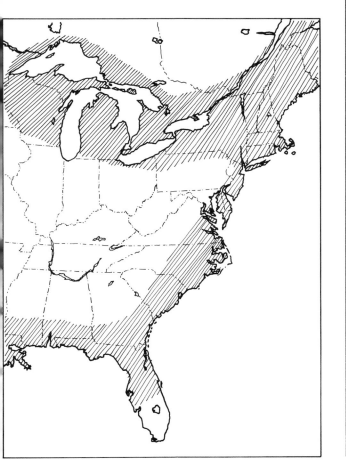

Fig. 5–7. D. intermedia, *plant in flower. While the leaves resemble those of* D. anglica, *the blades are somewhat narrower and smaller, and there is more red pigment in the leaf tissue. One can see the longer stem formation, even in this early summer plant.*

FLOWERING SEASON: June to August.
TRAP SEASON: Spring and summer. Forms hibernacula.

Fig. 5–8. D. intermedia *leaf from shade-grown plant (so red pigment of sun-grown plants would not interfere with photographic contrast). Note the longer-stalked peripheral, or trapping, glands and the almost sessile central digestive glands. Several stalked glands are bent over the remnants of small prey.*

DESCRIPTION.—This species is unusual in that the plant stems are quite long. In fact, the plant can reach a height of up to 20 cm as the season progresses. The left has much the same form as in *D. anglica*, except that the trapping blade is shorter and narrower (0.5 by 1.0 cm) and has a diffuse, dark red color when growing in the open. The flowers are usually white.

GENERAL.—This species has the largest range of any species of *Drosera* in the eastern part of North America. In the southeastern coastal plain, the species reaches its greatest size, and often the whole plant is deep red. It is regularly seen in more wet areas, particularly on the margins of streams, ponds, and drainage ditches, where it will grow into the water and sometimes over the surface in dense mats. The phenomenon of vegetative apomixis can be observed frequently in *D. intermedia*. (See Chapter 2, p. 21, for a discussion of vegetative apomixis.)

Drosera filiformis Raf.

BOTANICAL NAMES: *Drosera filiformis* Raf. Two forms or varieties are also generally recognized within the species: *D. filiformis* v. *filiformis* Raf. (also known as *D. filiformis* v. *typica* Wynne) and *D. filiformis* v. *tracyi* (Macfar.) Diels.
COMMON NAMES: Threadleaf sundew, dew-thread.
RANGE: As a species, from Cape Cod along the coastal plain into southern Mississippi.

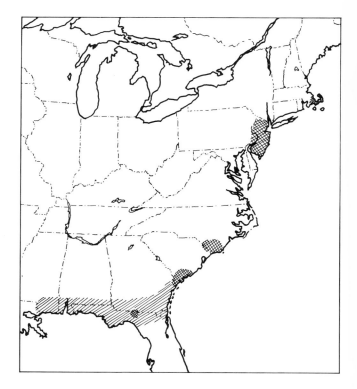

FLOWERING SEASON: June.
TRAP SEASON: Spring and summer. Forms hibernacula.

DESCRIPTION.—The leaves are erect with short petioles and filiform gland-bearing leaf blades. In *D. filiformis* v. *filiformis* (typica), the leaf blades measure to 25 cm and have bright red to purple glands; in *D. filiformis* v. *tracyi* they measure up to 50 cm and are uniformly green. The hibernacula of v. *filiformis* (typica) are by far the more hirsute. The flowers, which are always rose pink, are the largest of any variety of *Drosera* and are borne on tall scapes.

There are some floral differences between the varieties. The flowers of v. *filiformis* (typica) measure up to 1.5 cm across, and the outer margins of the petals are smooth. On observing the flower with a hand lens, one will note that the two anther lobes are joined at the tip and that the stamen filament is pale red. In v. *tracyi*, the flower is even larger, up to 2 cm across, and the petals frequently have somewhat scalloped outside margins. The larger anther lobes are separate, and the stamen filament is green.

GENERAL.—This is our largest sundew, and a stand of them growing in a savannah or on the margin of a bog is truly an impressive sight, especially on a dewy morning with the sun shining through the plants.

The smaller, red form occurs from Cape Cod south into the New Jersey Pine Barrens, where it is plentiful, and it is found in disjunct locations in the eastern Carolinas, Georgia, and a recently described location

Fig. 5–9. D. filiformis *v.* tracyi. *A stand of the plants in a closely cropped savannah, the glow caused by the early morning sun striking their glandular leaves.*

Fig. 5–10. D. filiformis *v.* tracyi, *a single plant. The color is diffusely pale green.*

Fig. 5–11. *Leaves of the two varieties of* D. filiformis, *v.* tracyi *being pale green and v.* filiformis (typica) *having red glands.*

65 / *The Sundews*

in northern Florida. The larger, green form grows in the southern Gulf coastal area, where it is very common. The ranges of the two forms reportedly overlap in South Carolina. In spite of this small area of sympatry, the two forms have not been found in the same stand, and a natural hybrid is not reported, although hybrids have been produced in the greenhouse.

As you will have noted, there is a minor problem

Fig. 5–12. D. filiformis *v.* filiformis (typica). *A stand of the species in eastern North Carolina. This variety has a red cast, which can be appreciated even at this distance.*

Fig. 5–13. D. filiformis *v.* filiformis (typica), *a single plant in the New Jersey Pine Barrens. The glandular character of the leaves is clear, and they are generally reddish.*

Fig. 5–14. *Flowers of* D. filiformis *v.* filiformis (typica). *These are always rose pink, and the species has the largest flowers of any* Drosera *in North America.*

Fig. 5–15. *A hibernaculum of* D. filiformis *v.* filiformis (typica) *just breaking in the early spring.*

Drosera brevifolia Pursh.

BOTANICAL NAME: *Drosera brevifolia* Pursh. Presently unacceptable synonyms or additional related species: *Drosera annua*, *Drosera leucantha*.

COMMON NAMES: Shortleaf sundew, dwarf sundew.

RANGE: In suitable locations throughout the southeastern United States, more commonly in the Gulf coastal plain.

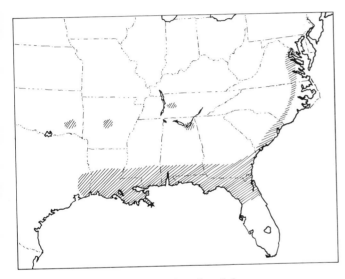

FLOWERING SEASON: April to May.

TRAP SEASON: In good locations, the leaves remain over winter. However, in locations that dry in the summer, the adult plants may die off, and seeds will germinate the following year. There are no winter hibernacula.

DESCRIPTION.—This is a tiny rosette measuring barely 2 cm across at the most. Larger plants occur toward the west, in Louisiana and Texas. The leaves are prostrate, and the trap blades taper back nearly to the rosette center. The petioles are extremely short to nonexistent—hence the specific name, *brevifolia* (meaning short leaf). The end of the blade is somewhat flattened so that the leaf is cuneate (wedge-shaped). The entire plant is frequently red-purple when growing in the open. The flower scape is uniquely gland-bearing, and the flowers are pink or white.

GENERAL.—A small nomenclatural flurry occurred in the sixties when a botanist attempted to discredit the species name *brevifolia* and to divide the populations into at least two species, *D. annua* and *D. leucantha*. But his arguments were not at all clear, let alone persuasive or documented, and he made the error of relying heavily on flower and leaf color and a few other questionable and very minor differences that could have represented environmental influence. His thesis was followed by a far more reasoned and well-researched paper by Carroll Wood urging the preservation of *brevifolia* as a species name. Wood's study

Fig. 5–18. *Flower of* D. brevifolia. *Note the glandular scape, which readily distinguishes this species from several others.*

Fig. 5–19. D. brevifolia, *the smallest species of* Drosera, *has a very short petiole and a wedge-shaped leaf with a flattened end.*

of the problem took four years to complete, and, during that time, the suggested substitute names began to appear in the literature.

This is our tiniest sundew, and one must look closely even to find individuals and then differentiate them from *D. capillaris*, with which they often grow. *D. brevifolia* is often present in glittering masses on a roadside bank or sandy flat. Closer inspection of such areas discloses the individual plants. *D. brevifolia* seems to prefer a drier habitat than most of our other species of *Drosera*.

DIFFICULT IDENTIFICATIONS

I think the biggest difficulty most people have is the differentiation of *D. capillaris* from *D. rotundifolia*. First of all, forget the common error of flower color. Second, *D. rotundifolia* is found in more northern and mountainous regions, whereas *D. capillaris* is located in the southeastern coastal plain (except for a few bogs in piedmont Carolina, where I have found the two together). Third, and most important, the trap portion of the leaf of *D. rotundifolia* is most often wider than long, but occasionally perfectly round in younger plants or in new spring leaves, while the blade of *D. capillaris* is longer than wide.

D. brevifolia can be confused with the above two species, especially when they are all seedlings. The dwarf sundew has a very short petiole and a wedge-shaped leaf blade tapering nearly back to the rosette center. The easiest identification marker for the beginner is the scape of *D. brevifolia*, which is gland-bearing, whereas the other two species with which it may be confused have smooth scapes. Also the purple-red color of the plants of *D. brevifolia* is distinctive from the lighter red of *D. capillaris*, but the difference is subtle and requires experience to discern.

D. anglica and *D. intermedia* do have some range overlap, but this is confined to a narrow area in Michigan. The former is more northern and has a very short stem and wider and longer leaf blades, which are green with bright red glands. *D. intermedia* is more southern and has a much longer stem in older plants, a smaller leaf, and red pigment in the plant tissue when growing in the open.

By the way, a rare natural hybrid of *D. filiformis* v. *filiformis (typica)* and *D. intermedia*, which had been described in the New Jersey Pine Barrens, was recently rediscovered. Its appearance is intermediate between those of the two parents, and plants studied thus far in the greenhouse appear to be sterile. The hybrid is sometimes known as *D. x hybrida*. *D. rotundifolia* x *D. anglica* (*D. x obovata*) is not infrequently found in northern bogs where the two species are sympatric. The leaf form is intermediate between those of the two parents and may be difficult to distinguish from that of *D. anglica* unless the two are compared side by side. Finally, *D. rotundifolia* x *D. intermedia* has been reported in New Jersey, the usual rule of intermediacy causing the leaf blades to appear almost round.

VI. The Butterworts (*Pinguicula* L.)

The Genus

BOTANICAL NAME: *Pinguicula* L. Family Lentibulariaceae.

COMMON NAME: Butterwort.

RANGE: Members of the genus occur all across the northern half of North America and extend down into the coastal plain in the southeast and into northern California in the west.

FLOWERING SEASON: Varies with the species; generally, early spring into early summer.

TRAP SEASON: The plants native to southern areas retain their leaves over winter; those native to northern and western areas form hibernacula (winter buds).

DESCRIPTION.—The plant is a rosette with stalkless leaves, the older ones lying prostrate, and the younger ones nearly so. The thin leaves are elongate and narrow more or less to a blunt point at the free end. They are generally flat on the main surface, but the margins are variably rolled in different species. The surface of the leaf is studded with nearly microscopic sessile (stalkless) glands, which impart a fine, pebbled texture. The leaf feels greasy to the touch—hence the name, from the Latin word *pinguis*, meaning fat, and the suffix *-ula*, meaning little one. The plant is pale yellow-green in most species, but reddish in one. The roots are fibrous and brittle, extending 2–6 cm into the ground.

The flower scape is often gland-bearing, is 4–20 cm long, and supports the single flower at the top. Multiple scapes appear successively during the early growing season. The flower is zygomorphic, and the corolla is at least partly sympetalous (i.e., the petals are fused together at their bases). The three lower petals form a lower lip with or without partial division into lobes, and the two upper petals form the upper lip. Near its base, the corolla narrows into a tube of cylindrical shape, and this terminates in a spur, which is even narrower and of variable length. Originating from the inner surface of the lower lip and at least partially covering the entrance to the corolla tube is a slightly bulging structure, called a palate, which is bearded

Fig. 6–1. Pinguicula vulgaris. *These plants are showing the formation of early autumn hibernacula in their centers. The general plant structure is evident.*

Fig. 6–2. *The flower of* P. vulgaris.

(hairy). This may be exserted (projecting outward beyond the flared corolla surface) or not, depending on the species. The hairs on and around the palate and in the tube of each species are seen to have a distinctive structure when they are examined under the microscope.

The stamens and pistil have a unique arrangement characteristic of the family. They are located deep in the tube. The stamen filaments are thick, stocky, and curved, and each is capped by a spherical anther with yellowish pollen. There are two stamens placed next to each other in the same plane, so that the curvature of the filaments nearly causes the anthers to touch. Behind the bases of the stamens and attached to the receptacle is the spherical ovary with a very short style. The stigma is bilobed and modified. The posterior lobe is greatly reduced; the anterior lobe is flattened and somewhat elongated, so that it hangs over the anthers like a veil or apron. The upper surface of the stigma is covered with a sticky material to hold pollen and support its germination.

The flower tends to encourage cross-pollination rather than selfing. The pollinator extends its forebody deep into the corolla tube and, as it does so, deposits any pollen from another flower on the anterior stigma lobe. As the pollinator withdraws, it tends to lift the apronlike anterior stigma lobe upward (because of friction and a tight fit), and expose the two anthers, so that it may pick up pollen from that flower without depositing it on the upper surface of the anterior stigma lobe, which has been turned back against the wall of the corolla tube. (See drawing at right.)

The trapping mechanism is quite simple. Small prey landing or crawling on the upper surface of the leaf become mired down in the glandular secretions and are held fast until digestion and absorption take place. During the active trapping season, the rolled edges of

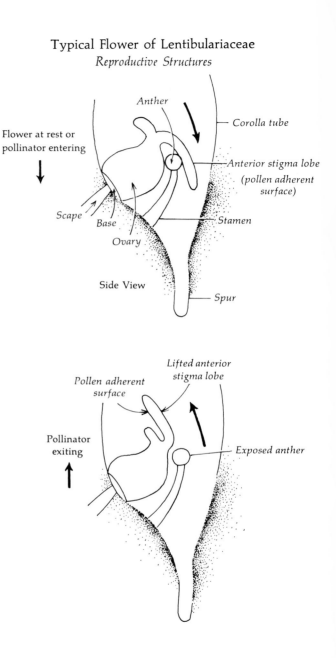

Typical Flower of Lentibulariaceae
Reproductive Structures

Anther

Flower at rest or pollinator entering

Corolla tube

Anterior stigma lobe
(pollen adherent surface)

Scape

Base

Stamen

Ovary

Side View

Spur

Pollinator exiting

Pollen adherent surface

Lifted anterior stigma lobe

Exposed anther

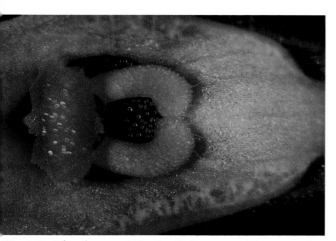

Fig. 6–3. *The flower of P. planifolia, opened by removing the lower lip. One can clearly see the apronlike lobe of the anterior stigma overhanging the partially exposed anthers atop the two curved, stout stamen filaments in the tube. Near the right side of the picture is a portion of the spur.*

the leaf tend to curl in some species, but seldom to the extent or with sufficient speed that they actually engulf the prey, as has been suggested in the past.

Studies by scanning electron microscopy have recently shown that the *Pinguicula* leaf has two kinds of glands on the upper surface, these being randomly intermingled. There are stalked glands that perhaps are more important in capture, and sessile glands that possibly are more active in digestion. Absorption probably takes place at the bases of the digestive glands.

GENERAL.—You will note the poverty of common names for these little plants as you read along. This clearly indicates the small regard and interest that has been accorded *Pinguicula,* except in a few scholarly instances. The plants are rather inconspicuous outside the flowering season, and there is a loose resemblance to several other herbs growing in the same habitat. In many instances, the new naturalist will have difficulty spotting *Pinguicula* right away. In late spring, however, the colorful little flowers, atop their tall, fragile scapes, nod in the breeze, and the plants are then much more obvious, especially when massed. The color of the flowers and the patterns of venation, if any, are characteristic for most species, and we will stress them. Until one gains experience, the vegetative part of the plant looks very much the same from species to species.

Not only are the plants small, vegetatively inconspicuous, and perhaps not as interesting to most people as *Sarracenia, Dionaea,* and *Drosera,* but many people are repelled by the cool, greasy feel of the leaves. This texture is due to the rather fragile, almost weakly succulent nature of the leaf tissue, in addition to the glandular secretions on the upper surface.

I know of one minor economic use for *Pinguicula.* People in the countries of northern Europe have mixed the leaves or leaf extracts of certain species with milk in order to curdle it and prepare a pudding dish much like junket or yogurt. Beyond this, the plants will have to be accepted and studied on their own merits. And they are fascinating plants.

Natural or artificial hybrids of the southeastern species have not been found. My own preliminary experiments with many crosses show that seeds are produced but do not germinate. There are reports of hybrids between the species of *Pinguicula* that form hibernacula, but these are mainly found in other countries.

The Species

Pinguicula vulgaris L.

BOTANICAL NAME: *Pinguicula vulgaris* L. There is a likelihood that a proposed split of what is commonly known as *P. vulgaris* in the west will eventually be accepted, resulting in another species, *P. macroceras* Link, with three varieties: v. *macroceras*, v. *microceras* (Cham.) Casper, and v. *nortensis* (unpublished at press time). See discussion below and map of outlines of ranges.

COMMON NAME: Butterwort.

RANGE: Northern boreal (subarctic) region south to the Great Lakes and northern California.

FLOWERING SEASON: June to August.

TRAP SEASON: Forms winter hibernacula, frequently with smaller off-budding basal hibernacula, or gemmae, which will grow into young plants in the spring.

DESCRIPTION.—In the northeastern range, the rosettes measure up to 11 cm. The leaves are pale yellow-green and are rather wide, with somewhat irregular margins and minimal marginal curling. The flower is violet with pale to white patches toward the tube. The corolla averages 1.1 cm across, and the upper lip is smaller than the lower. The beard is rather weak, being a confluent grouping of whitish hairs in the throat of the tube, and it is not exserted. The spur averages 5–7 mm and is sometimes notched or double.

The consistent variation of plants in the west has prompted the concept of a separate species, which is strengthened by the fact that, in the areas of overlap in the far northwest, the plants tend to maintain their separate characteristics. Generally, the rosettes of the variant plants are smaller (5–9 cm), and the flowers larger (1.5–1.8 cm). The main diagnostic difference is in the spur, which is longer in the west (6–11 cm), and also in the lower lip lobes, which are larger and tend to overlap or at least touch. Flower coloring is about the same. This plant has been proposed as a separate species, *P. macroceras*, by some botanists. The picture is complicated by still further variation within the putative new species. In the Aleutians, there is a variant with a very short spur (1–3 cm), but with the larger corolla. This plant is further designated v. *microceras*, and the longer-spurred, more southern plant, v. *macroceras*. Then, in the far southern end of the western range, in Del Norte County, California, populations were found with the long spur, but with smaller and well-separated corolla lobes more like those of the eastern *P. vulgaris*. The designation of the subspecies

Fig. 6–4. *Flower of plant from the "macroceras" range, Del Norte County, California.*

---

or variety *nortensis* (after the county) may be proposed for these populations, which have also now been found in immediately neighboring Oregon.

Japanese botanists, as well as many botanists in this country, have been mainly content to consider all these northern Pacific plants as varieties of *P. vulgaris*. However, there is a strong movement afoot to separate the plants as mentioned above. Unfortunately, data is far from complete, and claims for such separation are, at publication time, premature. Most of the original studies were done on herbarium material, which we are finding wholly inadequate for investigating the fragile *Pinguicula*. In fact, a monograph by a European botanist, who has studied the herbarium sheets and is promoting the new species designations, features a large number of photos of pressed plants and not more than a half-dozen live ones! Not to have studied the live plants more thoroughly is, in this day, totally inadequate.

Second, the all-important studies of reciprocal transplants *as well as* greenhouse studies in homogeneous environments have not been done. I cannot stress too

much the importance of such "behavorial" studies in investigating the variations of plants. They are too often overlooked by even the most experienced botanists, but they must be done no matter what the cost in time and frustration. Third, there are as yet not even elementary studies in genetic crossing. Finally, the chromosome number of typical eastern *P. vulgaris* is known to be 2n=64, but no chromosome counts have been done on a good cross section of plants throughout the western range, except for one plant of the proposed *nortensis* group, and this is 2n=32.

Eventually, after all these required studies are complete, the separation of the long-accepted *P. vulgaris* taxon may be acceptable without a doubt, and if so, the readers of this book will be prepared. But at the moment, there is far too much work to be done to be dogmatic on the issue.

GENERAL.—In the east, this species grows along the rocky and gravelly shores of lakes and streams in a thin layer of peat, or in and along the edges of both sphagnum and marl bogs, apparently able to get along well in several kinds of habitats. In the winter, one can often see hundreds of hibernacula and gemmae floating loose among the flotsam of lakeshore slack-water in quieter inlets. This is an apparently helpful mechanism for dispersal. In the west, the *"macroceras"* plants tend to prefer a narrower habitat; they grow in seeps on mossy outcrops of serpentine rock, often partially shaded, along the margins of small springs and bogs which often dry somewhat in the summer. In the far northwest, the plants are often seen on mossy sphagnum hummocks and in open gravelly seeps. Northern California plants with pale red leaves have occasionally been found recently growing interspersed among typical green-leafed forms.

Pinguicula pumila Michx.

BOTANICAL NAME: *Pinguicula pumila* Michx.
RANGE: The southeastern coastal plain from North Carolina into east Texas, but rare except in the Gulf area.

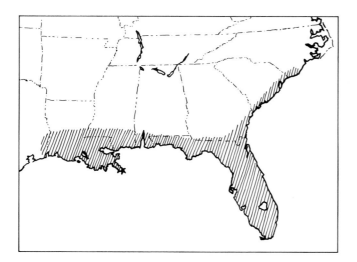

Fig. 6–5. *The flower of P. pumila. Note that the corolla does not appear fully expanded (a common occurrence in this small species) and that the yellow beard is not exserted.*

FLOWERING SEASON: April to May.
TRAP SEASON: No hibernacula are formed.

DESCRIPTION.—Most often, the rosettes are barely 1.5 cm across, but occasionally they reach 2 cm. The leaves are pale green and have pointed tips. The leaf edges are sharply rolled. The 1.0–1.5 cm flower is most often white but ranges to purple, yellow, or pink in Florida. The pale yellow beard is not exserted.

GENERAL.—This is our tiniest *Pinguicula,* barely discernible even when in flower. It tends to grow in moist, sandy places that dry somewhat—but not completely—during the summer.

Pinguicula lutea Walt.

BOTANICAL NAME: *Pinguicula lutea* Walt.
COMMON NAME: Yellow butterwort (referring to the color of the flower).
RANGE: The southeastern coastal plain from North Carolina to Louisiana.

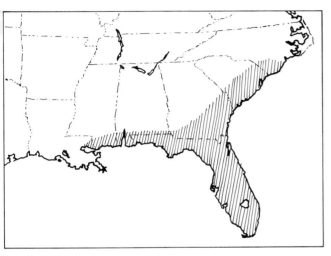

FLOWERING SEASON: February to May.
TRAP SEASON: No winter hibernacula.

DESCRIPTION.—These are pale green rosettes ranging 5–15 cm across, the larger ones tending to occur further south. The leaf edges are sharply rolled, and the leaf tips are pointed. The large (2.5–3.5 cm) flower is a brilliant yellow. This is the only large species in our area with a regularly yellow flower.

GENERAL.—A stand of *P. lutea* in flower in the earliest spring is a pretty sight and will not easily be missed. These plants grow in open, damp, sandy places that are shaded over to some degree in later summer by taller herbs and grasses.

Fig. 6–6. P. lutea. *The bright yellow flower is distinctive.*

Pinguicula caerulea Walt.

BOTANICAL NAME: *Pinguicula caerulea* Walt.
COMMON NAME: Violet butterwort (referring to the color of the flower).
RANGE: The southeastern coastal plain from North Carolina to the middle panhandle of Florida.

Pinguicula planifolia Chapm.

BOTANICAL NAME: *Pinguicula planifolia* Chapm.
RANGE: The gulf coastal plain from the mid-Florida panhandle west to Louisiana.
FLOWERING SEASON: March to April.
TRAP SEASON: No winter hibernacula.

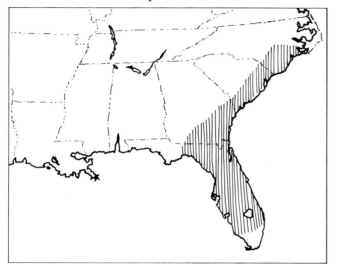

FLOWERING SEASON: February to May.
TRAP SEASON: No winter hibernacula.

DESCRIPTION.—The rosettes are pale green and measure 5–10 cm across. The leaves have sharply rolled edges and are pointed. The scapes are tall (to 20 cm), and the large 2.5–3.0 cm corollas are violet with prominent venation of a deeper violet. The palate beard is exserted and colored greenish yellow to cream.

GENERAL.—In the vegetative stages of later summer, it is practically impossible to tell this species from *P. lutea*, with which it sometimes grows. But in the spring, the deep-veined, violet flowers of *P. caerulea* certainly contrast with the yellow flowers of the other species.

Fig. 6–7. *The flower of* P. caerulea. *Venation is very prominent in this large flower. You can just see the top of the anterior stigma lobe down in the tube entrance.*

Fig. 6–8. *The flower of* P. planifolia, *with prominent exserted beard, pale purple color, and deeply incised corolla lobes.*

DESCRIPTION.—The large rosettes measure to 15 cm across, and the older leaves are flat and have only slightly rolled edges. The leaves are mostly a dull red to purple, although green races are common. The tall scape bears a 3 cm, violet, unveined flower, which tends to be darker colored around the tube entrance. A prominent characteristic is that the lobes of the corolla are deeply incised to at least half their length, so that, at a glance, it appears as if the corolla has ten lobes instead of five. The palate beard is exserted and bright yellow.

GENERAL.—In contrast to the previously described species of the southeastern coastal plain, this and the next two species grow in constantly wet areas, sometimes completely submerged in water for periods of time. The distinctly purple, unveined flower of *P. planifolia*, with its deeply incised corolla lobes, makes for easy identification in the spring. When the plant is not in flower, remember that it is the only species with reddish leaves. But different races and shade-grown plants of *P. planifolia* have green leaves, and in such cases there will be some difficulty in distinguishing non-flowering plants from the next two species.

Fig. 6–9. P. planifolia. *The leaves have a reddish color.*

Pinguicula primuliflora Wood & Godfrey

BOTANICAL NAME: *Pinguicula primuliflora* Wood & Godfrey.
RANGE: Gulf coastal plain from the western Florida panhandle into southern Mississippi.

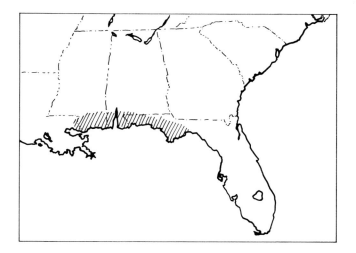

FLOWERING SEASON: February to April.
TRAP SEASON: No winter hibernacula.

DESCRIPTION.—The green rosettes grow up to 15 cm across, and the leaf edges are moderately rolled. The tall scape supports a 2.5–3.0 cm, very pale blue to violet flower with a white ring around the tube entrance. The external surface of the tube and spur is bright yellow. The palate beard is yellow and exserted.

GENERAL.—This species also prefers wet areas. It is often found in and on the edges of slowly moving streams, where it grows on hummocks of sphagnum, mostly in the shade. The flower is unique, as are the flowers of all the southern species.

There is another peculiarity of this plant that is shared to a far lesser degree by the other two principally Gulf coastal species, *P. planifolia* and *P. ionantha*. Small vegetative buds frequently sprout from the tips of older leaves in late summer. These can be seen growing from decaying leaves or ringed with other plantlets of varying ages around a larger "mother" plant.

Fig. 6–11. P. primuliflora. *There is a new plant bud arising from the end of an older leaf.*

Fig. 6–10. *The flower of* P. primuliflora. *The corolla is rose pink to violet with a white center.*

Pinguicula ionantha Godfrey

BOTANICAL NAME: *Pinguicula ionantha* Godfrey.
RANGE: A very narrow range in the mid-panhandle of western Florida.
FLOWERING SEASON: February to April.
TRAP SEASON: No winter hibernacula.

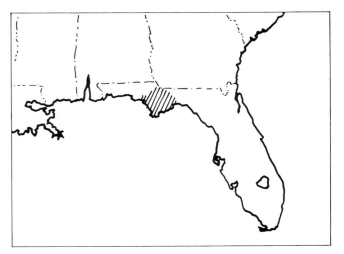

DESCRIPTION.—The flat, pale green rosettes measure up to 15 cm across. There is minimal rolling of the somewhat irregular leaf margins. The tall scape supports a 2 cm white to extremely pale violet flower with a ring of deeper violet around the tube entrance. The spur and external tube are olive to dull yellow. The palate beard is bright yellow and exserted.

GENERAL.—This species grows in very wet peaty or sandy places and in shallow water, often accompanied by *P. planifolia*. It has an extremely small range and is therefore somewhat endangered.

Fig. 6–12. *The flower of* P. ionantha. *This flower is white to very pale violet, with a darker ring in the center.*

Pinguicula villosa L.

BOTANICAL NAME: *Pinguicula villosa* L.
RANGE: North American boreal (subarctic and arctic) region from Alaska across extreme northern Canada.

FLOWERING SEASON: June to August.
TRAP SEASON: Forms winter hibernacula.

DESCRIPTION.—*P. villosa* is a small rosette, barely 2–3 cm across, with pale green, somewhat obovate leaves. The scape characteristically has numerous long plant hairs over the lower portion; hence the specific epithet. The flower, which is about 1 cm across, is colored pale blue to violet (with a white form recorded) and has fine yellow venation and a weak beard. The spur averages 0.5 cm and is somewhat conical.

GENERAL.—This plant grows on sphagnum tufts and hummocks in the extreme northern part of the continent, where it is well adapted to the subarctic climate. We have not been able to cultivate this plant for protracted periods, probably because it requires cool temperatures and some approximation of the long, cool summer days and the dark, very frigid winters of *P. villosa*'s natural habitat.

DIFFICULT IDENTIFICATIONS

Certainly when they are in flower—and sometimes when they are out of flower—there will be little difficulty in separating these species. Reference to the photographs will show floral differences at a glance. When the plants are considered regionally, identification is often more simplified.

In the area of the northeastern border between the United States and Canada, there is only *P. vulgaris*. Our western readers will have to contend with the "*macroceras*" hypothesis, if it is eventually established. Boreal botanists may run into *P. vulgaris*, the "*macroceras*" subgroups, and *P. villosa*. When the plants are in flower, identification will be easy, and the rounder leaves and hairy scape of *P. villosa* are also characteristic of that species.

On the Atlantic coastal plain of the Carolinas, there are only *P. lutea*, *P. caerulea*, and *P. pumila*, all with very distinctive flowers. Even out of flower, a stocky, tiny rosette is most likely *P. pumila*, which is generally uncommon in this area. Look for a maturing seed capsule to be sure that the plant is not a juvenile of one of the other two species! *P. lutea* and *P. caerulea* are common.

In the Gulf coastal region, especially upper Florida, there is the widest range of species, and identification by flower will be necessary in most instances, except when the red leaves of *P. planifolia* and the peripheral buddings of *P. primuliflora* are present.

VII. The Bladderworts (*Utricularia* L.) Family Lentibulariaceae

INTRODUCTION

We will depart from the format of the preceding chapters for several reasons. First of all, the taxonomy, biology, and range demarcations of the bladderworts are not as well known and understood as those of the other genera of carnivorous plants. There are no detailed botanical monographs on the American species, although at least one is known to be forthcoming. Second, the various species of *Utricularia* excite far less popular interest than the larger terrestrial carnivorous species. Indeed, if it were not for their annual flower displays, these rather minute plants could easily be passed over entirely by the novice, even though they are far more ubiquitous than any of the other carnivorous genera. There is hardly a salubriously boggy place that does not support at least one, and more likely several, species of *Utricularia*. Clearly, the genus requires and deserves far more study and attention.

In this chapter we will not have range maps, and there will be only representative photos of a broad cross section. Descriptions will be far more brief, mentioning a few salient points that should be sufficient to differentiate closely related species to the extent that we understand them at present. These will serve for easiest and most rapid identification. Many other points of difference will be omitted for the sake of clarity and to eliminate excessive technicality and controversy at this point, although we recognize their importance to the serious botanist. For these latter readers we suggest consulting the few available references.

The interested reader will find that a hand lens of good quality or even a microscope eyepiece is required for close examination and identification of many of these plants.

GENERAL DESCRIPTION OF THE GENUS

The species of *Utricularia* inhabiting North America are either aquatic or terrestrial, the former found as strands or mats of plants floating in quiet, acid ponds and bog-associated waters. The terrestrials grow most commonly in damp, sandy, acid soils, with the main parts of the plants at or below ground level. Without minute examination of the soil in the latter case, only annual flowering signals the plant's presence. The terrestrials also grow in sphagnum mats and hummocks, which are sometimes quite wet, and they can commonly be found in marl bog situations. Most of the species, aquatic and terrestrial, are frequently found in sand and peat muck or sphagnum slurries, an intermediate habitat. This capability of a biphasic habitat is an advantage for survival, since ponds with primarily aquatic species often partially dry out in the summer, and many areas where terrestrials grow flood during rainy seasons.

The form of the plant is a rootless, branching, green or brown stem 0.1–3.0 mm in thickness, from which arise whorls of finer green branches, which are sometimes divided and almost feathery, and which usually bear tiny bulbous traps. Properly, the traps themselves are the leaves, but no one quite knows what to call the trapless branchings which sometimes project up out of the soil or water in leaflike fashion. These have been called leaves or photosynthetic organs or branches, each name adequately implying the supposed function. The rootless, branching stem may reach a length of 3 m, or even more in the case of some aquatics.

The traps or bladders have a bulbous form, and they range in size from 0.5 to 3.0 mm according to age and species. They are attached to the finer branchings by a narrow stalk at the bottom of the bladder.

As mentioned above, many of the terrestrial species have narrow, flattened, pointed, green leaflike structures that have almost the appearance of seedling grass-blades. These arise from the stem or the base of the scape and project 1–5 mm above ground level. They apparently have a photosynthetic function.

Since *Utricularia* and *Pinguicula* are members of the same family, their flower structures are basically quite similar, so we will point out only some specific charac-teristics of the flowers of *Utricularia*. Reference to the discussion of the flowers of *Pinguicula*, pp. 71–72, will be helpful.

The aerial flowers of *Utricularia* are borne on an often tall, narrow scape, and they number from one to fifteen or even more, depending on the species. The scape may have along its course and at the branchings of pedicels one or several minute, leaflike appendages called bracts, bracteoles, or associated scales. The number, color, form, and location of these structures can be important taxonomically.

The flower itself is zygomorphic (two-lipped), as in *Pinguicula*, but it tends to be more flattened, and the size and shape of the spur are more variable. In *Utricularia*, the palate more completely obstructs the tube, and is hairless. There are fine red or brown reticulate lines on the palates of several species.

Fig. 7–1. U. macrorhiza, *showing the habit of the stem with whorls of branches bearing many small, nodular traps.*

Fig. 7–2. U. fibrosa, *a magnified view showing several traps.*

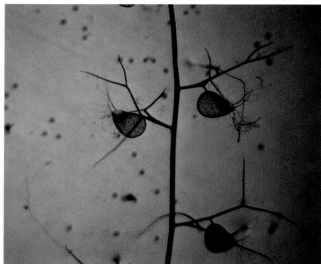

The flowers of several species of *Utricularia* can exist in two states, and the factors governing the appearance of one or the other are not known. A typically opened and expanded flower is described as chasmogamous. The second type, which is barely opened from the bud stage, and in which the two lips are still folded with only the spur protruding, is referred to as closed, or cleistogamous. Generally, cleistogamous flowers appear earlier in the spring.

Sometimes the presence of these two flower forms in the same species is confusing to the beginner. In the past, several professional botanists made errors and sometimes named the two forms as separate species when there was only one. Interestingly, even cleistogamous flowers, which pollinators cannot enter (or so we think at present), produce capsules of good, viable seed, as do the chasmogamous flowers of the same species grown indoors, away from potential pollinators. Since we mentioned in Chapter 6 that by design the typical flower of this family encourages cross-pollination, we have a thorny problem to solve in discovering how unpollinated flowers of many species of *Utricularia* (and of some species of *Pinguicula*!) still produce seed.

In the autumn, aquatic species of *Utricularia* often form a type of hibernaculum (winter bud) called a turion, and this either floats or sinks to the bottom of the stream or pool. In the spring, the green, nodular hibernaculum expands and then grows into new stems and branches. The turion originates from the growing tip of a branch or main stem.

TRAP FUNCTION

There is such a long, tedious history behind the ultimate disclosure of how bladderwort traps function that we will not detail it here. Suffice it to say that it reads like a who's who of early botany. (See general references, Lloyd, 1942.) The small size of the trap made early students reluctant to accept the fact that the mechanism of *Utricularia* could be so complex and rapid. We will present a very brief résumé of modern theory as to how the trap functions.

Near one end of the trap is a small opening surrounded by numerous plant hairs that are often branched and multicellular. The opening is guarded by a larger, upper-hinged veil of plant tissue called a door, and this is further supported in function by a smaller veil of tissue, the velum, which rests below the door on a thickened threshold. (See drawing, p. 84.) The door is weakly sealed against casual entrance by a thin layer of mucilage. There are minute glands on the surface of the trap, and located on the interior walls are peculiar quadrifid (four-pointed) glands.

During the resting stage, much of the fluid inside the bulbous trap is slowly absorbed, probably by the

---

Fig. 7–3. *A greatly magnified single trap of* U. gibba.

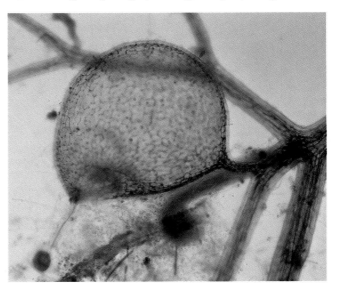

quadrifid glands. As a result, the water pressure outside the trap exceeds that within, and the sides of the trap appear pinched in, or concave. There has thus developed a negative, or suction, pressure within the trap which is now set to be sprung.

The trap is brought into action when a small aquatic animal brushes by one of the sensitive trigger hairs around the opening, or when the trap is otherwise severely disturbed. Stimulation of the hairs apparently releases an electrical action potential that in turn causes relaxation of the velum and thus frees the larger door to suddenly flip back into the interior of the trap because of the suction force developed in the resting stage. The opening and suction then allows an inrush of water along with the hapless prey, after which the door promptly closes. Since the door is hinged only one way, there is no escape.

Over a period of fifteen to thirty minutes, the trap resets by absorbing water into its interior and again recreating a negative suction pressure within. After a period of several days, the prey is digested as a result of enzyme activity that has been demonstrated in bladder extracts. The enzymes most likely originate from the quadrifid glands. If the prey is small enough for

Utricularia Trap Structure and Function

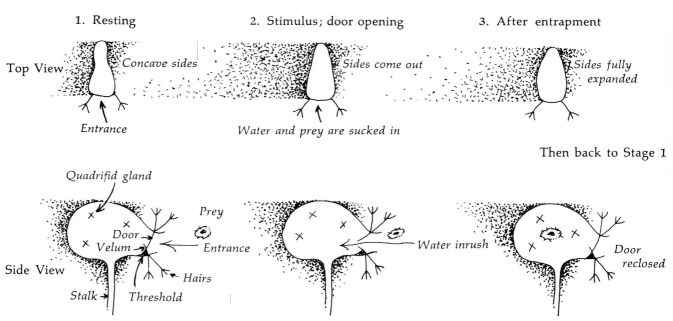

1. Resting

2. Stimulus; door opening

3. After entrapment

Top View — Concave sides — Sides come out — Sides fully expanded

Entrance — Water and prey are sucked in

Then back to Stage 1

Quadrifid gland

Prey

Door — Velum — Entrance — Water inrush — Door reclosed

Side View — Hairs

Stalk — Threshold

sufficient negative pressure to build up while it is in the trap, it is likely that another animal will be trapped before the one that was caught earlier is completely digested.

The speed of the trapping process has been estimated at 1/460 of a second and is far too fast to be slowed by the most advanced slow-motion cinephotographic techniques. When a mass of *Utricularia* is lifted from the water, one can often hear a fine crackling sound as the traps are sprung and air bubbles pop into them.

Generally, the prey of *Utricularia* is quite small, as one would expect considering the size of the trap. Most catches are minute water insects, protozoans, tiny crustaceans, rotifers, etc. The small waterweed *Wolffia* (duckweed) has been found entrapped in larger bladders of *U. macrorhiza vulgaris*, apparently ingested when the trap was accidentally sprung and not because the bladderwort had gone vegetarian. The bladderworts are capable of ingesting large numbers of mosquito larvae, which is of some universal interest to man, since the quantities ingested are apparently limited only by the number of traps available. Some species have even ingested small tadpoles.

There is a trick to how the bladder ingests prey longer than its longest dimension. Needless to say, it is a gradual process. If the tail of a mosquito larva is caught first and the animal is thus held fast, the flexible door closes around the protruding body of the larva and is still capable of effecting a seal tight enough to allow the absorption of water inside the bladder. Then the tail part is digested. If another stimulus occurs, the reset trap is then able to ingest the remainder or at least an additional length of the body until the whole animal is ultimately consumed in successive steps. The process is very roughly analogous to the spasmodic swallowing efforts of a snake ingesting a large prey.

The Species

For purposes of simplicity and easy identification, we will classify the species of *Utricularia* broadly by flower color, and then subclassify according to other easily observable characteristics that are reasonably consistent. Professional botanists frequently abhor such a system, since it is not "natural"; that is, it does not express real or theorized evolutionary relationships—relationships that are themselves often based on only the most tenuous and ephemeral evidence. However, nontechnical systems for the identification of complex genera are often most useful to the beginner, who may later progress to theoretical considerations if he so desires.

As you read the descriptions below, you may feel hopelessly mired in the great similarity of many of the species. But if you have at hand a plant to be identified and are able to make observations as you read, you will find that the system becomes workable with some experience. This is not a classical outline, but a linear key, which the reader must follow from beginning to end with any plant. For the convenience of the reader, the major categories are listed below. With the plant in hand or in sight, read from the beginning, eliminating inappropriate categories until you come to one that fits the plant. Then refer to the proper page number for detailed discussions of the plant or plants in that category.

1. Species with white flowers (p. 88).
2. Species with purple flowers (pp. 89–90).
   A. Aquatic.
   B. Terrestrial.
3. Species with mainly yellow flowers (pp. 90–95).
   A. Flowers with fimbriate bracts and sepals.
   B. Terrestrial; nonfimbriate bracts or bracteoles.

C. Flower scapes with radial floats.

D. Aquatic plants whose pedicels arch or recurve when in fruit.

E. Mainly aquatic plants with occasional subterranean branches and stretches of stem with traps alternating with filamentous "leaves."

F. Plants with threadlike stems tangled in mats and floating in shallow water.

The paragraphs below are not meant to be read and absorbed, but to be scanned as you look for characteristics of a plant in hand.

1. Species with white flowers. Only one American species, *Utricularia olivacea* Wright ex Griseb, *regularly* has white flowers. This aquatic plant is considered by many to be the smallest by weight of any flowering plant in the world. It has extremely slender, threadlike stems with bladders less than 1 mm across. *U. olivacea* floats in warm, mainly acid pond waters among algae and other species of *Utricularia* and is quite inconspicuous when not flowering. The species is occasionally found in ponds in the New Jersey Pine Barrens and in the coastal plain from eastern North Carolina into Florida. The flower, only 2 mm long and borne on a 2.0–2.5 cm scape, appears from mid-September to late October and, because of its small size, it can be easily overlooked. In fact, recent observations indicate that *U. olivacea* is probably more plentiful and widespread than previously supposed.

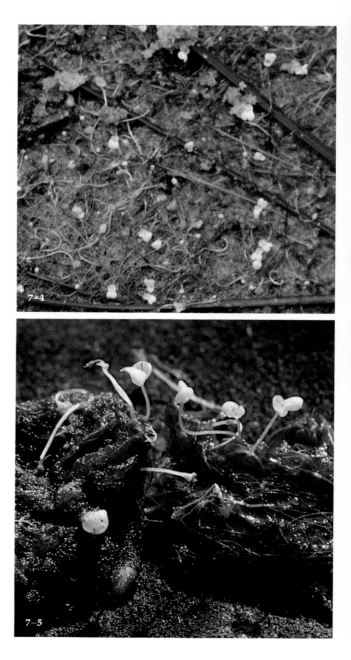

Fig. 7–4. *A small stand of* U. olivacea, *the tiny white flowers barely visible above the surface of the sand. Compare size with the pine needles in the picture.*

Fig. 7–5. *The white flowers of* U. olivacea, *our smallest bladderwort.*

2. Species with purple flowers. We will deal with one aquatic and two terrestrial species.

A. Aquatic.—*Utricularia purpurea* Walt. is very commonly found in suitable acid ponds or slow streams over the entire eastern half of the continent, particularly in the southern coastal plain. The long brown stems float submerged and give rise to whorls of branches which bear traps. The new spring growth tip often has a bright red color. Masses of plants are frequently quite large, and from May to September the purple flowers frequently cover the surface of the water with their bright blooms. The flowers are 1.0–1.3 cm and are borne singly on scapes which rise 7–10 cm above the water. The lower lip of the flower has two lobes, which are saccate (pouchlike). The part of the scape at water level has a slight fusiform (tapered at each end) swelling about 5 mm long, which has never been satisfactorily explained. It may possibly be an actual or primordial flotation mechanism to help keep the flower upright and out of the water. A white-flowered variant of the species, still with saccate lobes of the lower lip, has recently been found in a pond in New Hampshire.

B. Terrestrial.—While these two species may be observed in flooded areas, they are basically terrestrial, as is indicated by the fact that their vegetative portions are mainly subterranean. *Utricularia amethystina* St. Hil. (sometimes called *U. standleyae*) is found only in low, moist pinewoods of Florida and is rather rare. It is characterized by a whorl of minute, bladelike leaves around the scape at ground level. There is no information on the usual flowering period, and the flower is occasionally described as white, pale yellow, or whitish purple, these apparently being variants.

Fig. 7–6. *A stand of* U. purpurea, *the bright purple flowers distinctive above the surface of the water.*

Fig. 7–7. *A flower of* U. purpurea. *The saccate lobes of the lower lip can be seen clearly.*

Utricularia resupinata Greene occurs in very wet or mucky areas in mud flats and along the edges of lakes and ponds in southeastern Canada and the northeastern United States, and then skips an area until one reaches South Carolina and Florida. The plant frequently grows in a half centimeter or so of water, but with the main stem on the muck surface or below. There are narrow, small "leaves" frequently buried in the mud. The flower appears from May to September and is borne singly on a very thin, 2.5–12.0 cm scape. Using your hand lens, you will note that the paired bracts of the scape are joined, resulting in a tubular structure. The lower lip of the flower is not saccate, as it is in *U. purpurea*.

3. Species with mainly yellow flowers. Here is where we enter a very difficult area, since most of the North American species are yellow, and some of them are quite similar. There is still active discussion as to whether several of these related species are not actually variants of the same species. We will use the most widely accepted classification in this presentation.

A. Flowers with fimbriate bracts and sepals.—There is one species, *Utricularia simulans* Pilger (sometimes called *U. fimbriata*), found flowering throughout the year in low, moist pinelands of Florida. There are from one to seven 7–mm flowers on a slender scape, and the main characteristic is that the bracts and sepals have fimbriate (toothed or feathered) edges.

B. Terrestrial; nonfimbriate bracts or bracteoles.— We have three species. Remember that the habitats are subject to flood, but the bulk of the plant is firmly anchored in the ground, the scapes sprout from ground level, and all species have the tiny, grassblade-like "leaves."

   1. *Utricularia subulata* L. is a rather ubiquitous little plant occurring in suitably acid, sandy soils and bogs throughout the eastern third of the continent, but it is most prevalent in the southeastern coastal plain. The flowers number from three to seven on a wiry, zigzag, 7 cm scape, and they measure barely 8 mm across.

The lower lip is far larger than the upper, and the flattened spur is pressed against the back of the lower lip. This is the smallest yellow terrestrial species, and it flowers from May to November. Early flowers may be cleistogamous and pale yellow.

   2 and 3. *Utricularia cornuta* Michx. and *Utricularia juncea* Vahl are both larger species with prominent spurs projecting downward at an angle away from the lower lip of the flower. *U. cornuta* occurs in acid, moist, sandy places and bogs from Minnesota east to Nova Scotia, then down the eastern coastal plain, and west again to east Texas. *U. juncea* is a more southern species occurring mainly from the New Jersey Pine Barrens south into the coastal plain. Both flower from June to September generally, although *U. cornuta* tends to flower earlier in a location where both species occur together. At first glance, these species resemble one another, but the following differences are diagnostic: *U. cornuta* has a yellow-green scape which grows up to 30 cm tall and three to five chasmogamous flowers that measure up to 2 cm long with spurs 7.5–13.8 mm long. *U. juncea* frequently has a purple-green scape averaging 15–20 cm tall. The smaller flowers may be either cleistogamous or chasmogamous and measure about 1.0–1.5 cm long in the chasmogamous state, with spurs measuring only 0.7–2.4 cm long.

C. Flower scapes with radial floats.—These are aquatic plants, and there are two species, *Utricularia inflata* Walt. and *Utricularia radiata* Small. These plants occur in acid ponds and slow streams or ditches in the eastern coastal plain and occasionally inland to Indiana. They flower from May to November. The striking characteristic of both these species is a flotation device on the midpoint of the scape, consisting of air-filled arms radiating out on the surface of the water like the spokes of a rimless wheel. There are generally from four to

7-9 7-10

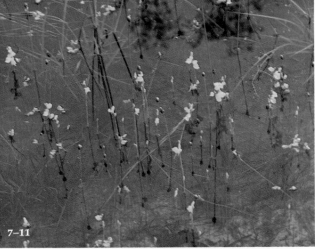

7-11

Fig. 7–8. *A stand of* U. cornuta *filling a bay (moist depression) in Brunswick Co., N.C. These were readily seen from a passing car; if they had not been flowering, one would hardly have suspected that they were there.*

Fig. 7–9. *An inflorescence of* U. cornuta *showing its large, bright yellow-gold flowers with long vertical spurs.*

Fig. 7–10. U. subulata, *a small terrestrial that is very common in its range.*

Fig. 7–11. *A stand of* U. juncea. *Even though the plants appear to be standing in water, you should note that they are firmly anchored in the ground and have simply been flooded. Closely inspecting and lifting a few plants is often necessary to distinguish terrestrials from aquatics.*

Fig. 7–12. U. juncea, *photographed at about the same distance as* U. cornuta *(fig. 7–9). The flowers are smaller and some are cleistogamous. The lower portion of the scape is purple.*

Fig. 7–13. *A fimbriate bract on the scape of yellow-flowered* U. simulans.

7-12 7-13

91 / *The Bladderworts*

Fig. 7–14. *A mass of* U. inflata *floating in the water of a roadside ditch in Columbus Co., N.C.*

Fig. 7–15. U. radiata, *a species similar to* U. inflata *but much smaller and with fewer "spokes" on the float.*

Fig. 7–16. *A single scape of* U. inflata. *The radiate flotation apparatus can be seen clearly.*

ten spokes. The arms are divided at the ends, where they frequently bear traps. These floats support the flowering part of the scape out of the water, while the lower part of the scape is below water level and continuous with the vegetative portion of the plant deeper in the water. The flowers number from three to seven on the average, but there can be up to fourteen in *U. inflata*. While there are several differences between these very similar species (some botanists would still consider *radiata* only a form of *U. inflata*), the easiest to use for identification is size: *U. inflata* has larger flowers and a much larger flotation apparatus, the float measuring to 25 cm across, with five to ten spokes; *U. radiata* is far smaller, the float measuring 6–8 cm across with four to seven spokes. *U. inflata* also has the interesting characteristic of producing tubers at the ends of some aquatic branches. Differences in flower spurs, which have been mentioned as differential characteristics in the past, are not reliable here.

D. Aquatic plants whose pedicels arch or recurve when in fruit.—There are two species.

1. *Utricularia macrorhiza* Le Conte (sometimes called *U. vulgaris*, *U. australis*, or *U. macrorhiza* ssp. *vulgaris*) is a rank grower and has the largest bladders, measuring up to 3 or 4 mm across. Very often the bladders are dark red to black. The aquatic stems are robust and grow several meters long. The species occurs from Labrador to Alaska, down the west coast into bogs of the Pacific coastal range, and in suitable scattered locations throughout the prairie states and from the northeastern states to southern Virginia. The scapes can be 60 cm tall, and they bear twenty or more 1.5 cm flowers. The spur is well developed and hooked; the palate has brown lines on it. The flowering period is from May through September.

2. *Utricularia geminiscapa* Benj. is often referred to as a "smaller edition" of *U. macrorhiza*. This species is limited to southeastern Canada and the northeastern United States. It has smaller flowers measuring to 0.8 cm. In addition to its small size, the species differs in having a less-developed spur without a hook. *U. geminiscapa* flowers from June to September.

E. Mainly aquatic plants with occasional subterranean branches and stretches of stem with traps and filamentous "leaves" alternating.—There are two closely related species. Both are literally anchored by branches reaching into the bottoms of ponds, and in both, trap formation occurs in periodic stretches along the stem, many alternate areas being without traps. The stem sometimes creeps along the bottoms of ponds and up into sphagnum tufts, where it may be growing on the open surface and appear somewhat like a low moss. Both species flower from May to September, and they range from the northeastern quadrant of the United States into southeastern Canada, westward to Alaska, and down into bogs of the Pacific coastal range.

1. *Utricularia intermedia* Hayne has a flower scape 5–20 cm tall that bears three to five flowers up to 2 cm. The flowers of *U. intermedia* characteristically have a cylindrical spur more than half the length of and positioned acutely behind the lower lip.

2. *Utricularia ochroleuca* R. Hartman has a similarly long scape with two to ten flowers measuring about 1.5 cm. In contrast to *U. intermedia*, the spur is very short, pyramidal, and vertical.

F. Plants with threadlike stems tangled in mats and floating in shallow water.—Here we will deal with six species in two subgroups:

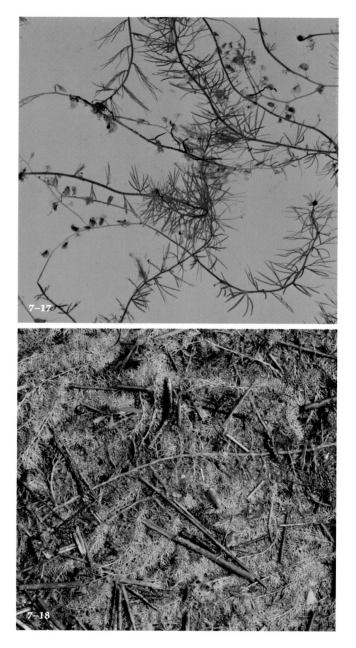

1. Lower lip about twice as long as upper, spur about half the length of the lower lip.—There is one species, *Utricularia minor* L. This plant can be found in bogs in the Pacific coastal range, in southeastern Canada, and in the northeastern United States. The scape is 5–15 cm tall and bears two to ten flowers that are about 1 cm across. The small spur is saccate and only 1–2 mm long. The plant flowers from May to September.

2. Lower lip about as long as upper, the upper lip not lobed.—The spur is almost as long as the lower lip. There are five species, and the morphology of the flowers is essentially very similar in all, the differences between species mainly centering on sizes and geographic ranges.

a. *Utricularia gibba* L. has the smallest flower of this last group, the scape being 3–7 cm tall and bearing one or two 0.5–1.0 cm flowers, which can be seen from June to September. The range is Pacific coastal bogs and ponds and the eastern half of the continent.

b. *Utricularia fibrosa* Walt. is about the largest species in this last group. The erect scapes are 10–40 cm tall and have up to seven 1.5–2.0 cm flowers. The species grows in the coastal plain from Massachusetts to Texas, and it flowers from May to November.

c. *Utricularia floridana* Nash is mainly confined to Florida and southeastern Georgia and is similar to *U. fibrosa* except that the scape is flexuous instead of erect, and the flowers number from ten to twenty-five but are only 1.2–1.5 cm wide.

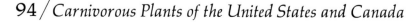

Fig. 7–17. *Stems of* U. intermedia. *You can see that the bladder-bearing branches alternate with foliar areas.*

Fig. 7–18. *A mass of* U. intermedia *that has crept up on a mudflat margin of Lake Michigan.*

d. *Utricularia foliosa* L. resembles *U. floridana* in some respects, but it ranges more widely, from Florida to Louisiana, and has flowers which measure up to 2 cm. One very fine distinction requires a good hand lens and necessitates cutting a cross section of the stem to examine the vascular bundles. This species has two sets of vascular bundles, while all other North American species have but one circular ring.

e. *Utricularia biflora* Lam. is very similar to *U. fibrosa* and occurs in much the same range. The difference in a shorter, 5–12 cm scape in *U. biflora*, and there are usually, but not always, only two flowers per scape—hence the species name.

7-20

Fig. 7–19. *The flower of* U. gibba *has much the same morphology as that of* U. fibrosa, *but the whole plant is smaller in all respects, the very fine threadlike stems often being mistaken for strands of algae.*

Fig. 7–20. *Flower of* U. fibrosa, *a large aquatic, with the upper and lower lips about equal.*

Fig. 7–21. *A single flower of* U. fibrosa. *Note the fine red reticulate markings of the palate.*

Fig. 7–22. U. biflora. *Very similar to* U. fibrosa, *except that the flowers are smaller and there are usually a pair on a short scape.*

7-21

7-19

7-22

95 / *The Bladderworts*

VIII. Growing North American Carnivorous Plants

Most of the carnivorous plants that have been discussed are not too difficult to grow successfully, given a few basic but rather strict requirements. There is clearly an increased interest in things botanical today, and there is special horticultural interest in unusual plants. Carnivorous plants now appear frequently in general houseplant catalogues, on the shelves of nurseries and commercial greenhouses, and in local discount and grocery stores. The number of dealers specializing in carnivorous plants is slowly but definitely growing.

What follows is a highly personalized version of my experience in growing carnivorous plants. I have successfully cultivated and propagated nearly all the species in this book, as well as numerous foreign carnivorous plants. I have found that in many cases several methods will work well, while in other instances one must adhere to a specific protocol. I do not pretend that my methods are the only useful ones, and the reader will very likely come across several other methods claiming equal success. You will do well to beware of brief and overly simplified instructions for a species inhabiting certain rather narrow ecologic situations. And by all means remember that any plant that has not been specifically bred and developed for the nursery trade—and this includes all our carnivorous plants—is never going to adapt perfectly in culture, no matter how apparently successful that culture may be. Plants always grow best where they are found wild, assuming that man or some other force is not destroying the surroundings to which the plant has adapted

during millenia of evolution. We can never hope to duplicate nature exactly, and certainly not by taking a few minutes to stuff a plant into a pot, taking it home to greenhouse or windowsill, and contentedly concluding, "Well, that's it."

This chapter is arranged in five sections. The first will deal with some general principles and definitions of the basic elements of the culture of carnivorous plants. Its main purpose is to be sure that we understand important terms and required growing conditions, and to lay out a broad overview of how North American carnivorous plants can be cultivated. Most of the information you will need is in this section. The second part will deal with specific genera and certain exceptional species within each genus; we will funnel the basic principles of the first part into the discussion of each genus, mentioning certain factors requiring emphasis in each case. There will be a third section on how you might manage or construct an outdoor bog. A fourth section will list a few specialized commercial mail order sources for carnivorous plants native to North America. Finally, we will say a few words about field collecting.

General Principles

DORMANCY.—We will start with the plants' rest period, which may seem an illogical beginning. But more growth failures occur from ignoring a plant's requirements for dormancy than from anything else. Not understanding and respecting this resting period has been the main reason for many desperate letters and phone calls to me from someone whose carnivorous plant is dying or just seems to be dying.

All temperate plants have a very definite dormancy period, some species of *Drosera*, *Pinguicula*, and *Utricularia* to the extent that they form winter hibernacula. There are also more subtle dormancy patterns in which the plants simply stop growing for some time each year. Dormancy is an evolved protective response to seasonal change that might otherwise injure an actively growing plant. Generally, this environmental change involves a period of decreased daylight, cooling, drying, or a limited combination of these conditions. Most dormancy in the temperate zones of North America will occur during the winter months, although a semi-dormancy may occur in some plants during late summer in response to drying, and we will discuss this later.

There are indications that dormancy is the result of the triggering of complex hormone mechanisms in plants. The mechanisms are not yet completely clarified and apparently vary with different species. In the future it may be possible to control dormancy in certain cultivated plants by various chemical and hormone applications. But right now the best thing to do about dormancy is to recognize it and help the plant to ride it out in the artificial conditions of cultivation.

A plant may indicate that it is going into dormancy in several ways: (1) A winter hibernaculum will form. Search for this in the autumn in appropriate species where the leaves seem to be failing. (2) The plant will simply stop growing after having put out leaves at a rather brisk pace. (3) Foliage will dry and die back, particularly in rhizomatous plants.

When dormancy begins, you must decrease watering to the point where the soil is *just* damp, and cool the plant by placing it for the winter in the basement, outdoors in a moderately protected area such as next to a building, or even in the refrigerator wrapped in a plastic or poly bag. (Do not freeze.) If you are growing plants under lights, reduce the photoperiod gradually to mimic the daylight period of the species' native habitat at that time of year. An almanac is helpful here.

If you try to "force" a dormant plant improperly with too much water and warmth, especially in a reduced photoperiod, you are inviting rot and the loss of the plant. A species in dormancy is not actively metabolizing and is less able to resist attack by harmful bacteria and fungi. Even when nurserymen claim to have forced any of the bulb plants, you will find that they have actually provided a managed and modified dormancy period first and have then exposed the plantings to all the conditions necessary for renewed growth.

WATER.—This would seem to be a ridiculous subject to spend words on, but it is important. There is water, and then there is water.

When growing carnivorous plants, one should use water that is as pure as possible. "Pureness" here refers to a low salts content and the lack of noxious or toxic waste materials. If you have access to an analysis, the total solids content of good water for carnivorous plants should be less than 50 parts per million (equivalent to 100 micromhos of electrical conductivity). The water should be neutral or acid rather than hard and alkaline. Ideally, such water comes from collections of rainwater, a granite bedrock well or spring (limestone-based wells often yield water with too many salts in it), distilled water, and *rarely* local tap water, which should ideally stand for twenty-four hours to allow the release of any chlorine. There is a new process and appropriate apparatus, called reverse osmosis, which produces pure water that is almost the equivalent of distilled, and it does so rapidly and cheaply and with less fuss. *Do not* use any of the various water softeners intended for "purifying" and softening water for home use. These usually exchange one set of toxic chemicals for another, "toxic" as far as plants are concerned. You may use certain more expensive double ion-exchange resin setups to desalt water. Be cautious of using water from local rivers and creeks, since most of these waterways are now heavily contaminated with toxic industrial and sewage wastes as well as with run-off of fertilizer from farmlands.

In general, during the active growing season, most terrestial carnivorous plants require even, constant moisture. The soil should definitely be wet to the touch, but not sopping and dripping.

HUMIDITY.—This factor is very important. Since nearly all our carnivorous plants are native to a bog of one sort or another, all require a good degree of humidity to grow well. In bogs and swamps, there is abundant surface moisture which evaporates and raises the relative humidity to saturation on warm days when plants are physiologically active. The plants present in a bog are adapted to such a situation and have come to require a good relative humidity to remain healthy. Most homes, especially when heated in the winter and air conditioned in the summer, are far too dry to support healthy growth if a carnivorous plant is placed casually on a coffee table or window-sill. Humidity will often have to be augmented in artificial growing conditions.

There are several ways to increase humidity for your plants. One is to grow them outdoors if you live in the eastern third of the country or on the Pacific slopes in the northwest, where there is abundant natural humidity during the warm seasons. Further inland, it is too dry. An ideal situation is to have access to a greenhouse where humidity can be carefully controlled. Another is to use a terrarium setup of pots enclosed in a glass case and lifted above water level. And here I will put in a word of caution: *Do not under any circumstances expose a closed small greenhouse or terrarium to direct sunlight for any protracted period.* In such a closed chamber, temperatures can and will rise precipitously to dangerously high levels, and this will kill plants in a short time.

A method of raising humidity in a room environment is to use one of the several efficient commercial humidifiers available on the market. Such equipment will often raise the humidity to such an extent that furniture will be ruined, so you will have to keep your plants in a water-safe area. Finally, the old saw about placing your pots of plants on a tray of moist pebbles does not really work at all well.

LIGHT.—Consistent with our thesis that plants respond best to environmental factors that exist where they evolved, sunlight is the best source of light for carnivorous plants. A minority of species (to be discussed later) require partial shading at some time during the growing cycle, but most do well in full sunlight. Natural light has the proper intensity and spectral composition for the best development and growth of plants.

Not everyone has access to a greenhouse, and perhaps local climate precludes keeping carnivorous plants outdoors. The increased use of various sources of artificial light has come about, mostly with excellent results. We recommend the use of fluorescent lights, rather than incandescent, because of the problems with temperature and spectral composition that arise with the latter. Equally effective are cool white fluorescent tubes and the more expensive but supposedly longer lasting special growth tubes. The best minimum fixture is one with a white or aluminized reflector and four 48-inch, 40-watt tubes which produce an intensely lit central area of usable size. Keep in mind that there is considerable loss of intensity at the ends of fluorescent tubes and out laterally from the fixture. The photoperiod is adjustable with available inexpensive automatic timers, so the grower may exactly duplicate seasonal changes. Fluorescent lights are cool, but they can be made even more so by removing the balast from the fixture, lengthening the wires, and then mounting the balast away from the growing area. Often, the minimal warmth of fluorescent light fixtures is desirable when growing plants in unheated basements. Humidity can be maintained with small tabletop humidifiers or with terrarium setups, which will be quite safe since, with fluorescent light, internal temperature buildup will not be a factor.

I have had considerable experience with fluorescent lights and still use them for handling certain seeds and cuttings that require precise adjustments of light intensity, photoperiod, temperature, and humidity. There is one major problem with the use of artificial lights, and this involves the drop-off of light intensity as lights are raised farther away from plant surfaces. The intensity of light varies inversely with the square of the distance of the source from the plant surfaces. To illustrate, if lights are set 20 cm from the plant surface and the grower then moves them up to 40 cm (twice the distance), the light intensity is not cut in half, but cut down to one-fourth. This is no problem when growing prostrate rosette forms such as *Dionaea*, most species of *Drosera*, *Pinguicula*, and some species of *Sarracenia*, because you can set the light at the optimal distance and leave it. But there is a problem with

growing the tall, erect pitcher plants and with maintaining flowers on taller scapes; both require lifting the light fixtures to a greater height as the plants grow. As one raises the light source to accommodate a rapidly growing erect *Sarracenia* pitcher, one is decreasing the light intensity to the growth crown near the soil surface where other buds will arise. The relative deficiency of light to the new leaf buds then results in etiolation—the production of long, leggy, soft green growth that falls over easily. These taller, leggier pitchers require even more raising of the light bank, further decreasing light intensity to newer buds, and the spiral continues until crown rot develops. Compounding the problem is the effect of the decreasing light on prostrate rosettes that the grower may have mixed in with the taller species.

The obvious solution is to have several light setups according to plant habitus and ultimate height. Very prostrate rosettes (*Dionaea*, most species of *Drosera*, *S. psittacina*, *Pinguicula*) under one setup, intermediates (*S. purpurea*, *Drosera filiformis typica*, etc.) under a second setup, and the tall, erect pitcher plants in some other sort of arrangement. I have never found a totally satisfactory solution to the problem of raising the tall species of *Sarracenia* under lights, but there is a partial, albeit expensive one. That is to use three light fixtures—one placed above and one placed at each side at an angle so that the light is directed toward the soil surface.

TEMPERATURE.—Most plants respond best to temperature variation in a twenty-four hour period. Let the temperature drop 5°–10°C at night. Temperature and humidity are intimately correlated, and light is an important indirect third part of this equation. Increased light frequently means rising temperatures and therefore higher humidity requirements. For example, well-lit plants kept in high relative humidity can tolerate warmer temperatures than poorly lit or drying plants. Likewise, good lighting and higher temperatures are going to require a higher input of humidity.

I am often asked about the maximum temperatures a plant or group of plants can survive without harm. Again, common sense refers us to the native environment, and again, one will find a good almanac very useful. During growth periods it would be nice to have a daily temperature of 30°–35°C with the optimum 5°C drop at night, but we are seldom so fortunate. Generally, plants that are native to the northern reaches are best kept at a temperature rarely exceeding 30°–32°C; plants indigenous to the southeast may endure temperatures up to 35°–37°C for short periods during the day. Even wider adaptation can occur in some cases if the change is brought about slowly.

One is more likely to find an adaptive plant among a batch of seedlings that in mature plants plucked from the wild, since in sexual reproduction a recombination of genetic factors occurs. One of these recombinations may be more adaptive to the new artificial environment, whereas it would have been selected against in the original habitat.

What about minimum temperatures? Here again, we can refer to the situation in nature. But my own experience has disclosed much more adaptability with lower temperatures than with higher. I have successfully grown most of the carnivorous plants discussed in this book outdoors in central North Carolina, where temperatures reach down to –18°C for short periods of time on several nonconsecutive winter nights a year. An intense, prolonged cold snap would be different. (I recall many an Ohio winter where, for days on end, the *high* for the day was freezing—0°C!) There are confirmed reports of healthy plants of southern *Sarracenia flava* and *Dionaea* surviving, reproducing, and of *S. flava* even hybridizing with native *S. purpurea* in Pennsylvania bogs year after year. There are unconfirmed reports of year-to-year survival of the same plants outdoors in an even more northern latitude in Michigan. Both these areas support a population of native carnivorous plants which would be expected to survive in local culture.

My conclusion would be to use these examples as guidelines. Northern outdoor growers should protect plants of southern origin during winters until experiments with one or two expendable plants prove otherwise. Use of the tub method of growing allows the removal of plants to protected areas over winter; and burying tubs and pots to the rim will often protect against severe root freezing. Covering growth crowns with a mulch over winter can also be helpful, but be sure to remove the mulch in the spring. Snow, by the way, is an excellent mulch.

POTTING.—If you are growing your plants in some sort of potting rather than in natural or homemade outdoor bogs, you will be confronted with two decisions: clay or plastic pots; drained or undrained ones.

For a long while it was thought that clay pots were inherently toxic to carnivorous plants, but now it has been shown that this is not so. Toxicity was due to the buildup of absorbed salts in the walls of older clay pots, especially pots that had previously been used for heavily fertilized plants. Salt buildup in a clay pot is disclosed by variegated, crusty rings of crystal material that do not wash off easily. You can use new pots, but eventually even the minimal quantities of salts in the soils of carnivorous plants will absorb and build up in the walls of the clay pots. At this point you would have to discard the pots, or soak and carefully clean each one—expensive and tedious processes.

For these reasons I suggest the use of plastic pots and tubs, with one or two exceptions to be noted later. Plastic pots do not absorb salts, and their surfaces are easily cleaned with good brushes. No soaking is required, except to remove unsightly but harmless stains, if desired.

Generally, it is better for the beginner to use pots or tubs with drain holes. More frequent watering will be required, but considerable experience and judgment are needed to use successfully the ultimately more carefree drainless system. Improperly managed pots with no drainage can result in rot and the eventual buildup of small amounts of salts in solution. I grow most of my specimens of *Sarracenia* without drainage. This requires that every year or two I spend some time filling the tubs to the brim with water, then tipping and emptying the excess water, repeating the process several times with each tub in order to dilute and wash out most of the salt buildup. I use drainage with most other terrestrial plants. One should also be cautious of using a pot-in-a-saucer-of-water setup; this amounts to an undrained system unless one remembers to flush water through the pots periodically.

If you do use undrained larger pots and tubs, you will find it helpful to plunge into the growing medium a smaller pot with drainage holes. The drain holes will allow water to rise up into this smaller pot, and you can use it as a monitor to check water levels. It is also useful to water your tubs through this smaller signal pot. Most of the smaller rosette plants do poorly if water is applied directly and regularly to their foliage. Indeed, sphagnum itself can turn brown and die in the area where even the purest water is applied daily. Water added to the tub through the signal pot will settle into the growing medium from beneath and will percolate out sideways and upwards towards the bases of the plants.

SOILS.—The soils for carnivorous plants should be loose, porous, poor in salts and nutrients, and highly acid—with one exception to be noted later. In my experience, live green sphagnum moss is the best growing medium for the culture of carnivorous plants.

There are several reasons for success with sphagnum. Many carnivorous plants grow in sphagnum bogs in the first place. Living and growing sphagnum naturally maintains acidity and a low level of salts and nutrients. Sphagnum has unique water-retaining properties and will maintain a proper level of moisture. At the same time, it is porous and allows rapid drain-off of excess water. Sphagnum has natural defense mechanisms against fungi and some algae. Finally, sphagnum is a good health indicator; the moss is quite sensitive to toxic materials, and if the sphagnum starts dying, the carnivorous plants will likely follow.

Sphagnum moss is actually a genus, *Sphagnum*, of about sixty species of varying colors and textures. We do not know of any documentation that the color indicates acid-producing capability, although the theory has been offered informally. Only experts on mosses are able to distinguish the species by name. For practical horticultural purposes, you will be interested in two kinds of sphagnum distinguishable by habitus: the coarse, rapidly growing species which are quite useful for tall, robust plants such as *Sarracenia*; and more compact, smaller, slower-growing sphagnums which should be used for small rosette plants that would soon be overwhelmed by the faster-growing sphagnums. One can chop coarser sphagnums up and use this as a medium for seedlings or small plants, but eventually a growth bud will take hold, and the sphagnum will resume its rapid growth, necessitating repotting.

By the way, sphagnum is the only "soil" we know of that grows for you. As the pots and tubs begin to brim and overflow with growing moss, you can trim it off and use the trimmings for more plantings.

Unfortunately, live sphagnum is often hard to find in many areas. Larger nurseries or plant shops may be able to help you. But do not buy either so-called sheet moss or milled sphagnum. In most instances, dealers will have only dried, dead, brown sphagnum in bags or bales. As long as it is not milled or fertilized, you can use this in the bottom layers of your pots and tubs, but you should still topdress with a layer of live green sphagnum. You can thereby compromise somewhat where live sphagnum is difficult to get but the dried, so-called long fiber sphagnum in bales is readily available.

My second choice is to use native sandy coastal plain soils, especially those from the Atlantic southeast. Be sure to collect it east of the clay banks but not too close to the ocean, where there may be salt contamination. Check with state agricultural authorities to be certain that it is permissible to transport soils from, within, or into a state from a particular location, which might be quarantined because of the suspicion that there are organisms of agricultural disease in the soil.

Of course, fewer people are going to be able to obtain coastal plain soil than live sphagnum. But you can make a pretty good substitute using the following formula:

   1 part fine, washed silica sand (Remember *not* to use that collected at the seashore—it will be too salty.)
   1 part fine peat (Use German or Canadian, in which there is no added fertilizer.)
   1 part small grade perlite

Mix this well and wet it down thoroughly; then let it stand for one or two weeks to age, during which time it will develop proper acidity and a balanced micro-flora.

The least desirable growing medium is plain sand. When using any sand—either plain or in the above formula—avoid lake bottom or river sand. These sands will have all the pollutants and nutrients that were present in their waters, unless you go to the trouble of a great deal of washing. Washing may not be successful anyway, since these sands often have clay particles which hold salts and toxins tenaciously. Fine white silica sand can be purchased by the bag in hardware stores, where it is sold for sandboxes and decorative purposes. However, it may have been collected near salt water, and if you have any doubt about this, a quick wash with running water will rid white sand of any salts.

FERTILIZATION.—After stressing the need for growing media free of nutrients and salts, this heading might seem confusing. But if you start with a near zero level of nutrients, and you know what you add (if anything), then you know exactly where you stand. While most carnivorous plants grow nicely in sphagnum year after year without prey or fertilization, they will become more robust and flower more if *very lightly* fertilized. The best way to avoid the decision of whether to chance fertilization is to put your plants outside on fine days so that they can periodically catch prey naturally. Very little actual prey will do. Feeding with meats is definitely not advisable. It is too easily overdone, and such highly concentrated nutrients result in plant damage and serve as a breeding ground for fungus infection.

The most useful fertilization schedule—and I would not start one unless you feel your plants are doing poorly because of a lack of nutrients—is a minimal monthly feeding during the growing season. Use any balanced fertilizer diluted to about ten times the dilution suggested by the manufacturer's label for houseplants. If the instructions say one teaspoonful per quart for houseplants, make it one teaspoon per ten quarts for carnivorous plants, or smaller total quantities in proportion so that the volume can be easily handled. I am partial to seaweed fertilizers rather than highly refined materials since these mixtures of natural origin supply many other essential elements in trace amounts. The method of application is a light, fine, once-over spray using any of the hand sprayers available. The material will be absorbed foliarly. It is not necessary to pour solutions into the pitchers of *Sarracenia*.

Remember, live sphagnum is a good health monitor. If the sphagnum dies and you have recently fertilized, you may have overdone it. Also, a growth of slimy, blue-green algae on top of the sphagnum indicates too much phosphate, and likely too much total fertilizer and salts. By the way, I have seen plantings in clay pots in which there was a nice ring of blue-green algae on top of the sphagnum near the pot edge only, indicating that salts were being leached from the old pot into the sphagnum next to the wall.

PESTS.—This refers to the pathogen variety, not inquisitive visitors, pets, and other people's kids with carelessly probing fingers. Yes, carnivorous plants are susceptible to insect pests, and some others. Plants grown by reliable nurseries should be free of pests, but field-collected plants may be another matter.

We have already mentioned the pests of *Sarracenia* in Chapter 3. The first step in ridding pitcher plants of *Exyra* is to seek out any adults resting on the inside pitcher walls just below the lip. Remove them with long forceps and kill them. Next, if you see telltale signs of larval activity, such as orange frass in the pitcher, a dry ring around the pitcher top, or a web spun across the pitcher mouth, cut off the afflicted pitchers at the very base and burn these leaves. The consecutively laid eggs hatch sequentially, so you may be trimming infested pitchers for several weeks after killing a fertile adult female moth.

The *Sarracenia* root borer, *Papaipema appassionata*, can be diagnosed by an enlarging conical pile of bright orange droppings collecting at the growth crown near the soil surface. If you are lucky, you can seek out the larva with narrow forceps, pull him out of his rhizome tunnel, and destroy him. If you cannot remove the larva, you can use a method recommended by J. A. Mazrimas which involves the instillation of a dilute solution of malathion into the larval tunnel with a medicine dropper. Dilute the malathion according to the instructions on the label. Your plant is likely to develop rot and die if destruction has been too great before the larva is detected and removed. Adults are no problem in *Papaipema*.

Other likely pests more peculiar to cultivation are scale and mealy bugs. Scale is an insect appearing as rows of soft tan to hard brown "turtleshells" averaging 2–4 mm across. They are often on the surfaces of the growth crown where they may not be seen and will often show up indirectly as deposits of black mildew on the upper pitchers. This is because the scale releases plant juices which in turn support mildew growth. Mealy bugs are fluffy, white, powdery insect colonies located on leaves. Both these insects can infest the interior of *Darlingtonia* pitchers, where they are very hard to diagnose and eradicate. The treatment is a light but complete spraying with malathion diluted according to manufacturer's instructions. The scales will not drop off when they die and dry up, but they are then harmless. Two weeks after the first spraying, repeat the treatment to deal with any recent egg hatchings. Malathion does not affect eggs, only the insects. Inspect all new plants entering your collection for these pests. They will spread rapidly and hide in all nooks and crannies of the plant's anatomy.

Fungi can be a problem with *Pinguicula* and especially with the terrestrial species of *Utricularia*, less often with *Drosera* and *Dionaea*, and rarely with *Sarracenia*. Fungus infections usually indicate that something is wrong with the culture: dormancy requirements are not being met properly, or something is wrong with the proportions of light, water, and temperature. Fungus infection is indicated by the moist, slimy browning of leaves with rotting of growth crowns, which tends to spread across the pots and tubs like a small plague. You may also see typical fuzzy growths of fungal fruiting bodies. The first step of treatment is to analyze and correct the culture problem that originally led to the fungus infection. The second is to apply a fungicide. There is now available a very effective and safe new systemic fungicide called Benlate or Benomyl. Make up a suspension according to the instructions on the label and apply two ways: first, give a good spraying to all leaves in the pot or tub; then uncap the sprayer and apply the solution to the growing medium as though you were watering. You will be effectively treating the infection on the leaves as well as in the whole plant systemically through the roots.

Finally, I would again remind you that the adults of the *Wyeomyia* mosquito larva often found in pitchers of *Sarracenia purpurea* are harmless to people and plants.

PROPAGATION.—This is a fitting finale for this section on general principles. It has been said, and rightly so, that you do not actually have a plant until you are able to propagate it through seed or vegetative means. There is sound basis for this statement. If the plant is doing well enough to reproduce, and the resulting seedlings and buddings in turn grow to maturity, then you are doing something right. We will discuss specific pollination, seed harvesting, and sowing with each genus. Here we will mention general procedures.

The best medium for seedlings is finely chopped, live green sphagnum mixed with an equal amount of fine, washed, white silica sand. Plain sand is second choice. Place some small granite pebbles in the bottom of a plastic 2–4 inch seedling pot with drain holes, then put in your soil mix and water thoroughly. Next, sow the seeds directly on the surface. Do not sow too thickly, and do not cover any carnivorous plant seeds.

As a general rule, seeds that mature in the spring or very early summer are ready for immediate sowing and will germinate promptly. Storage at ordinary room temperatures will result in deterioration. Seeds that ripen in the fall must usually undergo a period of damp cold treatment called stratification. This is best accomplished by placing the seed-sown pot upright in a poly bag, sealing it, and refrigerating it for either the whole winter (for growth in a greenhouse) or a minimum of six to eight weeks (for growth by artificial light).

Immediately after sowing spring-ripened seeds, and after stratifying fall seeds, place all pots in essentially the same growing conditions as adult plants, using slightly filtered rather than full sunlight. Germination of North American species will take place in two to four weeks. When watering, use a medicine dropper so as not to displace very tiny seeds. When seedlings appear, it is best to administer a dose of Benomyl or Benlate (diluted according to directions on the label), using a medicine dropper. This will help forestall, in many kinds of seedlings, a common fungus disease called "damping off."

Seeds of aquatic species of *Utricularia* can be sown

directly in the water in which they will be grown (see following section); those of terrestrials, on soil surfaces. In fact, terrestrial species of *Utricularia* and some other carnivorous plants will prove almost weedy because they self-seed so readily. Needless to say, this is a rather pleasant weediness.

Seeds, even the tiny spring seeds, can be safely stored. Allow the collected seeds to dry in the air for a day or two in a quiet place where air currents will not scatter them. Then either place the seeds in airtight, dry plastic or glass vials or wrap them in squares of waxed paper. Label them properly and keep them in the refrigerator. Under refrigeration, even the most evanescent seeds keep for at least three to five years.

In cases where it can be used, vegetative propagation is the quickest way to obtain larger plants of exactly the same characteristics as the parent plants. There are several basic techniques that can be used for North American carnivorous plants:

1. **Vegetative apomixis** can be used for propagation when it occurs in *Dionaea* or *Drosera intermedia*. (See Chapter 2, p. 21.) Carefully remove the plantlets from the scapes and plant them so they can take root.

2. **Natural leaf budding** will occur often in *Pinguicula primuliflora*, occasionally in other Gulf species of *Pinguicula*, and fairly commonly in many species of *Drosera*. When the buds are large enough and have developed root systems of their own, they can be separated and planted.

3. **Stolons** will send up plants in *Darlingtonia*, and these can be cut loose and replanted after they develop their own root systems.

4. **Rhizome branchings or buddings** of *Sarracenia* that occur naturally in successful culture may be separated by cutting after you are sure that roots are being produced on the rhizome branch.

5. **Rhizome cutting** may be used for the propagation of *Sarracenia*. I am grateful to Steve Clemesha of Australia for developing and telling me about the following technique. Uncover the upper half of a horizontal rhizome of a good-sized plant until the top of the stem is fully exposed, but leave the roots in the soil. Using a sharp, fresh, single-edged razor blade, very carefully slice perpendicularly into the rhizome about halfway. Repeat this at several points. Leave the top of the rhizome uncovered, and in a few weeks new growth buds will appear at the sites of the cuts. As soon as new roots develop, you can complete the separation of your new individual plants.

6. **Leaf cuttings** work very well with most species of *Drosera*, moderately well with nonbudding species of *Pinguicula*, and fairly well with *Dionaea*. The procedure is best done early in the growing season. The technique is to cut off a fairly fresh but mature leaf at the base of the petiole and to place the entire leaf right side up on a bed of moist, finely chopped green sphagnum. The lower side of the leaf must be flat on the surface of the sphagnum, and you can achieve this by pinning it with toothpicks or by spreading a single layer of coarse cheesecloth over mounded sphagnum. The plantlets will come up through the holes in the cloth. David Kutt originated this idea. Leaves of *Dionaea* and *Pinguicula* will sometimes do better if the end of the petiole is placed in the sphagnum rather than simply being laid on top. Try several cuttings both ways.

Next, place your pot of cuttings in a high-humidity chamber in the *shade* and in a warm—but not hot—place. The floor of the greenhouse under a bench is excellent in the spring. Over a period of several weeks you will note the appearance of plantlets from the margins and surfaces of the flattened-out leaves and at the petiole end of some leaves. Let these grow until you are certain that roots have formed. During this period the mother leaf will usually blacken and die. Watch out for fungal growth, and treat it appropriately if it appears. When the young plants are well rooted, transplant them carefully to separate pots and *slowly* acclimate them to proper lighting and decreased humidity over a period of several more weeks.

Genus Notes

Now that we have discussed some basic principles of growing carnivorous plants native to North America, we will mention each genus and relate important specifics, emphases, and exceptions.

DIONAEA.—*Dionaea* requires a complete period of dormancy as outlined in the previous section. When dormancy begins in cultivation, the leaves usually turn black and wither, but the rhizome remains healthy and fleshy as long as there is no attempt to force growth. Use a small variety of sphagnum or the coastal plain soil mix; drainage is best. When the plants are actively growing, there should be abundant light, humidity, and moderate but warm temperatures.

Dionaea's spring growth pattern begins with a few trap leaves that are not very large or well developed, followed by the flower scape if the plant is large enough to bloom. After flowering, larger and more typical leaves are produced all season long. Many will not wish to bother with the flower, and it can be cut off at an early stage, which tends to stimulate the formation of larger and earlier traps.

I have found that the best routine propagation method is by seed. You should have two or more plants in flower simultaneously, since selfing is very difficult because of the differing maturation periods of anthers and stigmas in the same flower. One can pluck individual stamens with fine forceps and touch these to the stigmas of another plant, but this is tedious. A quicker and more certain method is simply to bend the scapes and brush the open faces of two flowers on two different plants lightly against each other with a circular motion. This will result in a mutual transfer of adequate pollen from one to the other. The flowers of *Dionaea* open successively, and you will have to repeat the pollination process daily as new flowers open. The small, black seeds mature in six weeks and can be sown immediately as described in the previous section.

SARRACENIA.—This genus also requires very careful attention to dormancy. If you are growing the plants in an undrained tub, remember to cool the tub and cut way back on watering until the sphagnum is just damp. During the active growing season, give plenty of light, water, and moderate warmth, and grow the plants in green sphagnum. Daily misting of the foliage with water is also beneficial.

While most species of *Sarracenia* do well in full sun, *S. purpurea* requires a bit of special care. Give it full sun early in the spring to encourage flowering and good pitcher formation. (Flowers can be clipped off in bud if not desired.) Then, as the days of summer become hot, you should place the plant in partial shade and protect it from breezes. This species does better with somewhat higher humidity allowances than other members of the genus.

Leaf cuttings have never been regularly successful as a propagative method for *Sarracenia*. The separation of natural rhizome buds and branchings and the partial rhizome cutting technique described in the previous section work very well. Seed is not at all difficult if one is willing to follow instructions for stratification. If the plants are not outdoors, you will have to pollinate the flowers. Allow the flower to mature by being open three to five days. Then lift a petal and gather a small quantity of pollen from the umbrella cup with the flat edge of a toothpick. Dust the pollen over the small stigma lobes on the inside surface of the umbrella points. Only one lobe need actually be pollinated, but do two or three to be sure. You can self-pollinate or cross between plants. Seeds will set by autumn, the drying and enlarging capsule eventually splitting. Remove the entire capsule, open it completely, and separate the seeds from their attachments onto a sheet of paper.

The various species of *Sarracenia* hybridize very readily, and you may wish to try your hand with simultaneous flowers of two different species. Remember to label properly.

DARLINGTONIA.—I have found this to be the most difficult North American carnivorous plant to grow truly successfully. I am not referring to a plant that seems to barely hang on for a year or two or three, but one that year after year produces new, larger, and more vigorous growth with active stolon production. Unfortunately, this species is being offered more by general nurseries under the illusion that it is an easy plant to keep in all areas of the country. Some nurseries even display the plants, usually newly shipped-in adult specimens stuck into a pot of sphagnum. As the display plants succumb, they are promptly replaced by more from the refrigerator.

After much trial and too much error, I have found the following methods to be most successful for me. First of all, this is one plant that should be grown in a clay pot, a new and unsalted one, of course. The reason for using clay is that water seeps into and through the pot wall and then evaporates, thus cooling the pot and roots. Remember, this plant's native home is in some very cold-water western bogs. The growing medium must be particularly loose and very poor in solubles. You can use a mixture of live, green sphagnum of the coarse variety and very coarse perlite; or as I prefer, simply a potful of washed, coarse *granite* gravel with live sphagnum as a topdressing. Such a potting will encourage a good supply of oxygen to the roots, as well as maintain humidity and cooling in the root area in spite of excellent drainage. Rinse liberal quantities of cold water through the pot daily. Keep the plant in a cool, humid, semishaded place in summer. If you live in areas with very warm summers, and nighttime temperatures do not drop below 20°–22°C for many nights in succession, you will very likely lose your plant. Needless to say, a proper period of dormancy is required for good health.

If your plant is doing well, it will produce stolons after a year or two, and once the buds are well rooted, you can cut and separate them as new plants. Or if you are using a large clay pot, let it fill with a mass of pitchers.

Seeds are easily produced by selfing or cross-pollinating flowers that have been open three to five days. Lift a petal, with forceps pluck off a stamen or two, and brush the anthers over the stellate stigma beneath the "bell." Seeds will set by fall and can be harvested and germinated as in *Sarracenia*.

I re-emphasize that this species is not an easy long-term subject, and plants are expensive. But it is very attractive, and many will wish to try it.

DROSERA.—You should be able to grow and propagate this genus very easily in smaller sphagnums, chopped sphagnum, or coastal plain soil mix. Do not become worried over the species that form hibernacula in the fall. These include *D. linearis, D. anglica, D. rotundifolia, D. intermedia,* and both forms of *D. filiformis.* The hibernacula of *D. filiformis* v. *typica* seem especially discouraging at this stage, since they are well covered with a dense coat of black hairs that, when moist, make the whole thing look as though it has rotted. But careful inspection by separating some of the hairs discloses the tight bud of bright green primordial leaves. Remember to cool and dry out the plants to the point that they are barely damp for winter. During the growing season, give all species very good light for maximum coloration and development.

D. linearis, as you will recall, will not be readily available, and it occurs naturally in marl or alkaline bogs instead of acid situations. We have found, however, that it adapts to acid systems in cultivation. Another good soil formula for this particular species was developed by J. A. Mazrimas and consists of equal parts of vermiculite and fine, white silica sand. You may add some dolomite or not. This mix has much the same aggregate consistency as native marl soils, although it is not alkaline without the dolomite.

Propagation is by leaf cuttings or seeds. Actually, you will find that many species of *Drosera* self-seed, and little plantlets will crop up all over soil surfaces. If not pollinated by wind, insect, or man, most *Drosera* flowers automatically self-pollinate as they close at the end of the day. The seedpods ripen in four to six weeks, becoming plump and dark brown to black, and seeds can be collected and sown immediately in most cases. *D. linearis* and *D. anglica*, however, by virtue of being very northern plants with late seed setting, will do better with a period of stratification.

PINGUICULA.—One must remember that in their natural environments many species of *Pinguicula* are shaded over by taller grasses during the hot late summer, after having had full sunlight in the cooler springtime.

Generally, *P. pumila*, *P. lutea*, and *P. caerulea* need semishading and drying of the soil to bare dampness as summer progresses. During the winter, be particularly careful with watering. I have found that placing the plants on slight mounds in their soils is helpful. Give full light, and water rather generously in the spring when new leaves and flowers are actively growing. Try to avoid getting water on the leaves.

The Gulf coastal species, *P. primuliflora*, *P. ionantha*, and *P. planifolia*, grow in very wet habitats, but in culture such wetness predisposes the plants to fungus attacks during winter dormancy. Therefore, allow the medium of even these plants to dry to bare dampness in the winter. *P. primuliflora* does well in light shade as in its native habitat.

P. vulgaris (and the putative "*macroceras*" subspecies) form winter hibernacula, and you must be especially careful not to overwater or overheat these. In fact, the best policy is to take the buds up, dust them with sulfur, place them in a poly bag with a strand or two of damp sphagnum, and refrigerate them over winter. As you examine these winter buds, you will often note smaller offset buds, or gemmae, at their bases. These too will grow into smaller, young plants in the spring.

P. primuliflora will form leafbuds by itself each year, and the other Gulf coast species will occasionally. Leaf cuttings are successful in these species. You can obtain seeds by pollinating the flowers yourself. Open the flower by grasping the lobes of the upper lip with the fingers and tear down the lower lip with forceps, thus exposing the stamens and stigma. Grasp a stamen with your forceps and carefully pull it loose. You will see the rounded, yellow anther at the tip. Rub the anther over the overhanging part of the stigma lobe of the same or another flower until you see pollen coming off onto the stigma surface. Over a period of four to six weeks, the ovary will swell, turn brown, dry, and then open to reveal very fine black seeds. Sow immediately and treat preventively for damp-off.

UTRICULARIA.—The terrestrials and some of the semiaquatics are no problem at all, becoming almost weedy in the pots of other genera. They self-pollinate and self-seed to the extent that I have followed the movement of a species from one end of a greenhouse to the other with the air ventilator current. A pinch of soil in which they are growing can be placed in another pot, and it will soon fill with bladderwort plants. Some species, such as *U. fibrosa* and *U. gibba*, can be grown in sphagnum slurries. These are made by filling a flat tray half full with live sphagnum and half with water.

The obligatory aquatics and semiaquatics you may wish to grow in a pool are a different story. Sometimes balanced aquaria work out well, although these may not be acid enough, and some species of aquarium animals eat the *Utricularia* traps. The best way to grow these plants is to make a special pool that resembles their natural habitat. You can either use a large plastic tub or a child's wading pool sunk in the ground to the rim, or dig a hole and line it with a continuous sheet of polyethylene. Place about 5 cm of sandy peat in the bottom and add pure water to the top, along with a few strands of live sphagnum and some chips of cedar wood, if available. Let the pool age for one week. During this period the wood and peat will settle, and the water should become clear and coffee-colored. If you are familiar with soil-testing kits, check the pH (acidity) of the water. It should be pH 4–6. If the water is not acid enough, you can add more sphagnum and peat and let it age some more. If you are familiar with chemicals, diluted sulfuric acid can be added slowly,

while stirring, until an initial low pH is reached. The pH should not be lower than 4, however. Sulfuric acid, even when dilute, should be handled only by those experienced in using dangerous chemicals.

Next, add your plants or turions. Allow for full sunlight, and replace the water as it evaporates. Some algae may grow at first, but you will note that as the plants of *Utricularia* become established, the water will clear, and it will also seem to buffer at the correct pH.

Remember that the aquatics will often appear to die off in late summer or autumn as they form winter buds or hibernacula. New growth will begin in the spring.

Aquatics are propagated by simply breaking the stems and placing a piece in another tub.

An Outdoor Home Bog

If you live east of the Mississippi River or northwest of the Pacific coastal ranges where there is good natural humidity, you can make artificial outdoor bogs of whatever size and number you wish. After finding which plants will adapt, the only limitation will be low winter temperatures. Winter protection can be provided by covering the plants. We have mentioned several plants endemic to the southeast that have adapted to winters farther north, and all North American carnivorous plants should adapt to northwest coastal regions. Since the requirements of winter dormancy are automatically met in outdoor plantings, most species actually do better in such a setting than indoors.

First of all, select a good area for your bog. It should receive full sun for at least half the day, and morning sun is preferable. For ease of maintenance, it is best constructed away from deciduous trees and their leaf-fall area. There is no reason why a bog cannot be placed among other home plantings. You may have to consider some sort of fencing or other protection against curious dogs and cats. Finally, place the bog convenient to whatever source of pure water you use. Natural rainfall will do for most of the time, but you may have to supplement with watering during summer dry spells.

Now for the construction. You can use one or several of the larger plastic tubs available in variety stores, or you can use any size of the inexpensive children's plastic wading pools. (Nest three or four of these, since they are thin-walled.) Do not plan your bog too small; you will always need a larger one than you originally thought. I would therefore strongly suggest the wading pools. Drill a 0.5 cm hole in the side wall near the bottom of whatever vessel you use. Even though the object of the container is to retain water, a small drainage hole (which will be below ground when you are done) is helpful to drain off excess water slowly after heavy rain and to provide a slow flow of fresh water through the bog.

Next, dig a hole large enough to accommodate the pool or tub at your bog site. Place the vessel in all the way to its rim and fill in soil around the outside to support the sidewalls. The surface of your container should be level. You may wish to place natural stone around the edges to conceal the plastic rim.

Now you must fill your bog with an appropriate growing medium. Again I suggest live, green sphagnum, preferably of the small, tufted, compact variety so that you can have mixed plantings without being concerned about a coarse sphagnum species taking over the smaller plants. If this type of sphagnum is not available, use the coastal plains mix. If live sphagnum is available in limited quantities and you cannot find enough to fill your bog, use the coastal plain soil mix or dried, baled sphagnum (*not* milled) in the bottom and topdress with live sphagnum.

An interesting variation is to sink a smaller tub or

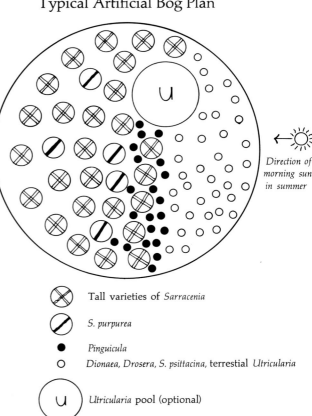

Typical Artificial Bog Plan

Direction of morning sun in summer

⊗ Tall varieties of *Sarracenia*

⊘ *S. purpurea*

● *Pinguicula*

○ *Dionaea, Drosera, S. psittacina,* terrestial *Utricularia*

U *Utricularia* pool (optional)

plastic container in your bog, this smaller one having no drainage hole. You can prepare this container as a pond for growing aquatic species of *Utricularia* as outlined in the previous section.

Water your new bog very thoroughly; this may take a few days if you are using dry sphagnum or soil as a base. In order to prevent hose water from hitting the leaves of *Drosera* and *Pinguicula*, you may wish to place a small plastic tube or two on the surface of your bog, with the tubes connected to a main hose so that the water will trickle in slowly over the soil surface. *Pinguicula* and *Drosera* do not seem bothered by rain pelting their leaves, as they are by artificial waterings on a regular basis.

When planting your bog, try to make the arrangements look natural. Do not line the plants up in neat rows but instead use circles, groupings, and curves. You must follow a broad basic pattern, however, if you wish mixed plantings. Place the tall pitcher plants together on one half so that they will not shade the other half of the bog from the morning sun. You may scatter *Sarracenia purpurea* among the tall pitchers—they will get plenty of light for good development early in the growing season, and later the tall pitchers will shade and protect them from hot sun and breezes. Also, plant *Pinguicula* on the edges of your pitcher plant stands, raised on little mounds 1–2 cm higher than the soil surface, and ranging in and out among the pitcher plants bordering the half of the bog reserved for smaller plants. This remaining open area can now be used for *Drosera*, *Utricularia*, *Dionaea*, and *S. psittacina*. A diagrammatic plan for such a bog is presented as an example.

Water your bog when natural rainfall fails and the tops of the sphagnum tufts seem to be drying. In late fall, trim off dead foliage (leaving pitcher plant phyllodia), and if you have severe winters, cover your bog with black poly sheets pinned down around the edges with stones. If you leave your bog uncovered, remove any deciduous tree leaves, particularly oak, as they fall on the surface. A large, moistened, flattened deciduous leaf will smother everything under it, including sphagnum moss, long before the leaf rots. Also, it has been shown recently that some leaves and other plant parts—again, particularly those of the oak—release chemical substances that inhibit or kill growing plants of species other than their own, a mechanism (allelopathy) that serves to reduce competition in natural settings and that will just as effectively reduce your bog to a shambles.

Some Commercial Sources of Carnivorous Plants

For the convenience of readers, I am listing several nurseries that specialize in carnivorous plants. There are many other dealers that sell one or two species, or that obtain their material from one of the sources below, but I have not included them since they are secondary and their lists are limited. Many dealers in wild flower plants also have for sale carnivorous species that are native to their areas.

I suggest writing several of the dealers to obtain their catalogues. Do not forget to enclose money for the cost of the catalogue, if necessary. Look the lists over carefully and compare them: prices and sizes of inventories vary greatly. Practically all have quality material and use proper packing and shipping procedures. Many offer interesting foreign species as well. The best time to order your plants is in very early spring. If offered, request optional special handling or air mail delivery.

When your plants arrive, open the package and inspect them immediately for quantity, quality, sizes, damage, and disease. Then plant your specimens promptly. This means preparing your planting area when you send off your order, not the day the package arrives. Incidentally, those who live close to any of these nurseries can visit them in person and purchase plants, thereby reducing delays and possible damage and confusion due to mailing. Hours are usually listed in the catalogue; if not, inquire.

Finally, this listing is intended for the convenience of the reader, and the author does not personally endorse or stand by any of the firms mentioned.

SUN DEW ENVIRONMENTS, P.O. Box 503, Kenmore Station, Boston, Massachusetts 02215. Catalogue—50¢.

ARMSTRONG ASSOCIATES, Box 94H, Kennebunk, Maine 04043. Catalogue—25¢.

ARTHUR E. ALLGROVE, North Wilmington, Massachusetts 01887. Catalogue—25¢.

NORTHROP'S INSECTIVOROUS PLANT FARM, P.O. Box 5, Hampstead, North Carolina 28443. Catalogue—25¢.

PETER AND PAM, P.O. Box 4415, San Fernando, California 91342. Free list.

INSECTIVOROUS PLANT ENVIRONMENTS, 26381 Whitman Street, Hayward, California 94544. Free list.

PETER PAULS NURSERIES, Darcey Road, Canandaigua, New York 14424. Catalogue—25¢.

Field Collecting

The basic rule should be: Do *not* collect from the field. I believe the exceptions to this concept are less than one per cent. True, many of the plants you may come across seem endless in a particular location or even in a very large geographic area. But equally true is the cliché that, if everyone who came through took some plants, there would ultimately be few or none left. Many states have laws intended to protect these and other valuable native plants, but the statutes are unenforceable and are all too often blatantly ignored. Small collections for serious research purposes are exceptions, and most states offer permits for this kind of collecting. Commercial firms are supposed to propagate their stock, but many have been known to collect regularly from the wild as orders accumulate. Such firms should be boycotted.

You may be fortunate enough to come into ownership of a natural bog, or to obtain permission to use a bog on someone else's property. The bog may or may not contain native carnivorous plants, but even if not, you can use it for careful plantings. If you are going to add non-native carnivorous plants to your bog, be sure to place them in a clearly marked-off area along one edge so that they will remain segregated.

You may have to do some work on the bog to prevent or partially reverse eutrophication. Cut out unwanted shrub and herbaceous growth from the center and from around the edges if the vegetation appears to be encroaching on true bog plants. Many bogs are partially drained in an ill-conceived attempt to try to make the land agriculturally useful. Block up any such efforts at drainage. Check the uphill slopes around a bog, especially water inflow areas, to be sure that sources of contamination and toxic materials are removed. Keep traffic in the bog to a minimum; much tramping about damages important bog plants and tends to create paths that soon become new drainage ditches. Ideally, you can even build a walkway around or across the bog, using creosote-soaked wood or abandoned rail ties, which will do only minimal harm to plants where the wood actually touches them. Such a walkway will reduce immensely the damage from traffic.

You may be able to stock your bog or build up your

private growing collection by taking advantage of dying or threatened stands of carnivorous plants. Many a bog on private property is undergoing rapid eutrophication, and perhaps the owner does not wish it preserved. Obtain permission to collect in such cases. Also, many savannahs and bogs are being drained and cleared for massive forestry and agriculture, some for the right of way for roads, and others are being flooded in dam projects. If you hear of such activity, get in touch with the current owner or project manager and obtain permission to rescue the plants.

Many times you will run across a doomed location that has far too many plants for your own collection, or perhaps has other rare and desirable noncarnivorous species as well. In cases of this sort, get in touch with local botanical gardens, especially those that are particularly interested in native plants. Some areas have native plant societies, and these people are always happy to go out on a "dig," as it is called, and will rescue and relocate endangered native plants by the truckload. In North Carolina, for example, the North Carolina Society for the Preservation of Wild Flowers has often worked in concert with the North Carolina Botanical Garden in Chapel Hill in various digs, relocating plants to the garden, which specializes in native flora, and to other native gardens in the state as well. This particular arrangement works out very well. But beware of some botanical gardens, even large ones. In spite of their prestige, many simply cannot care for carnivorous plants properly, and many donated collections have been promptly lost.

The best time to make field collections (again, do *not* collect except for serious research or for the sake of the plants) is very early spring when the plants are just budding, or in the autumn. Always collect a ball of the soil or sphagnum in which the plants are growing to help maintain them during transport. Be careful of damaging the tender leaves of *Drosera* and *Pinguicula*. Do not pile small tender plants but lay them carefully out in flat, shallow trays.

When you arrive at the transplant site with your plants, check them carefully for introduceable disease. If infested specimens of *Sarracenia* are found, trim and burn all affected pitchers. Plants infested with *Sarracenia* root borer are best destroyed. In all plants, cut off dying, damaged, or diseased leaves. It is also a good policy to wash the roots of larger plants clean in some place where the washings will not contaminate the new planting site. Then, plant as soon as possible.

Additional Reading

The following articles and books were selected to complement the material in this book. It is not a complete list of writings on carnivorous plants; additional information may be obtained from bibliographies in many of the scientific papers, from standard indices, or from librarians. Some of the entries are annotated as a guide. This list is a mixture of popular books and magazine articles as well as technical materials and journal papers. The nontechnical material is available through any good local library, directly or through interlibrary loan. The journal papers may be read in university or scientific libraries, or may also be obtained through interlibrary loan.

General

Argo, V. N. 1964. Insect trapping plants. *Natural History* 73:28-33. A brief introductory article with good photos.

Darwin, C. 1875. *Insectivorous plants*. New York: D. Appleton & Co. A classic work available in many libraries in a later American edition (1898).

Lloyd, F. E. 1942. *Carnivorous plants*. Waltham, Mass.: Chronica Botanica. Now out of print but possibly republishable, this book provides much information on anatomy, histology, and early experiments with digestion. There are many fine line drawings, but limited photos due to wartime conservation.

Poole, L. and Poole, G. 1963. *Insect-eating plants*. New York: Crowell. A good introductory book for young people; line drawings.

Rickett, H. W. 1966-1973. *Wild flowers of the United States*. Six regional vols. New York: New York Botanical Garden. This monumental work, featuring most of our native flowering plants in the regional volumes, has fairly good coverage of the carnivorous plants, although the information is brief and obviously somewhat scattered. One flaw is gross mislabeling of the *Drosera* photos. All color photos.

Rowland, J. T. 1975. Carnivorous seedplants: sources and references. *Hortscience* 10:112-14. A very good bibliography and reading list with many references to consult. A must for serious interest.

Shetler, S. G., and Montgomery, F. 1965. *Insectivorous plants*. Leaflet no. 447. Washington, D.C.: Smithsonian Institution. This free booklet features good narrative and some fine black and white photos. Some foreign species are mentioned and pictured. References.

Zahl, P. A. 1961. Plants that eat insects. *National Geographic* 119:642-59. A very fine color photo article featuring American species.

Venus' Flytrap
(Dionaea)

Affolter, J. M., and Olivo, R. F. 1975. Action potentials in Venus' flytraps: Long-term observations following capture of prey. *American Midland Naturalist* 93:443-45. This paper and the two that follow cover some aspects of electrical potential response in tissues of *Dionaea* and the physiology of trap closure.

Benolken, R. M., and Jacobson, S. L. 1970. Response properties of a sensory hair excised from Venus' flytrap. *Journal of General Physiology* 56:64-82.

Jacobson, S. L. 1965. Receptor response in Venus' flytrap. *Journal of General Physiology* 49:117-29.

Roberts, P. R., and Oosting, H. J. 1958. Responses of Venus' flytrap to factors involved in its endemism.

Ecological Monographs 28:193–218. A very complete paper discussing morphology and ecology; a must for serious interest.

Scala, J. et al. 1969. Digestive secretion of *Dionaea muscipula* (Venus' flytrap). *Plant Physiology* 44:367–71. An analysis of biochemical phases of prey digestion.

Pitcher Plants
(Sarracenia)

Bell, C. R. 1949. A cytotaxonomic study of the Sarraceniaceae of North America. *Journal of the Elisha Mitchell Scientific Society* 65:137–66. This paper and the two that follow comprise a set of classic papers that present good descriptions, field and natural history notes, and several fine black and white photos.

———. 1952. Natural hybrids of the genus *Sarracenia*. *Journal of the Mitchell Society* 68:55–80.

Bell, C. R., and Case, F. W. 1956. Natural hybrids in the genus *Sarracenia*, II: current notes on distribution. *Journal of the Mitchell Society* 72:142–52.

Case, F. W. 1956. Some Michigan records of *Sarracenia purpurea* forma *heterophylla*. *Rhodora* 58:203–7. A detailed account of the form of this species free of red pigment.

Jones, F. M. 1904. Pitcher-plant insects. *Entomological News* 15:14–17. This paper and the four that follow comprise a classic series describing various insects that themselves prey upon or live commensally with *Sarracenia*. There are excellent drawings and photos useful for identification, and the writing is clear and interesting—models for some of today's scientist-authors!

———. 1907. Pitcher-plant insects—II. *Entomological News* 18:412–20.

———. 1908. Pitcher-plant insects—III. *Entomological News* 19:150–56.

———. 1920. Another pitcher-plant insect. *Entomological News* 31:90–94.

———. 1921. Pitcher plants and their moths. *Natural History* 21:296–316.

McDaniel, S. 1971. The genus *Sarracenia*. *Bulletin of the Tall Timbers Research Station* (Tallahassee, Florida) 9:1–36. Does not offer anything more than the Bell, and Bell and Case papers above, but it is all in one place. Some taxonomic discrepancies, but good descriptions, range maps, and a fine set of line drawings of all species.

Mandossian, A. J. 1965. Plant associates of *Sarracenia purpurea* in acid and alkaline habitats. *Michigan Botanist* 4:107–14. This paper and the two that follow form an interesting and informative series detailing the author's studies in Michigan bogs.

———. 1966. Variations in the leaf of *Sarracenia purpurea*. *Michigan Botanist* 5:26–35.

———. 1966. Germination of seeds in *Sarracenia purpurea*. *Michigan Botanist* 5:67–79.

Plummer, G. L. 1963. Soils of the pitcher plant habitats in the Georgia coastal plain. *Ecology* 44:727–34. Soil structure and analyses of southeastern *Sarracenia* habitats.

Plummer, G. L., and Jackson, T. H. 1963. Bacterial activities within the sarcophagus of the insectivorous plant, *Sarracenia flava*. *American Midland Naturalist* 69:462–69. More study certainly needs to be done in this area of pitcher plant physiology and intra-pitcher ecology.

Plummer, G. L., and Kethley, J. B. 1964. Foliar absorption of amino acids, peptides and other nutrients by the pitcher plant, *Sarracenia flava*. *Botanical Gazette* 125:245–60. First reported use of modern methods for following absorption and transfer using radio-isotopes.

Wherry, E. T. 1929. Acidity relations of the Sarracenias. *Journal of the Washington Academy of Science* 19:379–90. In this paper the author first describes the *jonesii* pitcher plants of the Carolina mountains as a species. Contains interesting information on soil pH studies with respect to *Sarracenia*.

————. 1972. Notes on Sarracenia subspecies. *Castanea* 37:146–47. After years of controversy, this paper contains the author's decision to consider *Sarracenia jonesii* a subspecies of *S. rubra*. See also Bell (1949) and McDaniel (1971).

————. 1933. The geographic relations of *Sarracenia purpurea*. *Bartonia* (15):1–8. In this paper, Wherry makes his first presentation of the concept of two *purpurea* subspecies. See Wherry (1972) for his updated thoughts.

Walcott, M. V. 1935. *Illustrations of the North American pitcher plants*. Washington, D.C.: Smithsonian Institution. A limited edition folio of water color prints of all the species, long out of print but available in many libraries; a collector's item. Actually, the prints are limited because of the material selected for illustration, but they are beautifully executed. This volume is more important for the descriptive notes by Edgar T. Wherry and for a fine summary section (with bibliography) on insect associates by Frank Morton Jones.

Sundews
(*Drosera*)

Shinners, L. H. 1962. *Drosera* (Droseraceae) in the southeastern United States: An interim report. *Sida* 1:53–59. This and the following paper present both sides of the *D. brevifolia* controversy, which we mentioned briefly in Chapter 4.

Wood, C. E. 1966. On the identity of *Drosera brevifolia*. *Journal of the Arnold Arboretum.* 47:89–99.

————. 1955. Evidence for the hybrid origin of *Drosera anglica*. *Rhodora* 57:105–30. The author convincingly argues that hybridization of *D. linearis* and *D. rotundifolia*, followed by amphiploidy, is the origin of this species.

————. 1960. Droseraceae. *Journal of the Arnold Arboretum* 41:156–63. A brief summation with a key; also includes *Dionaea*.

Wynne, F. E. 1944. *Drosera* in eastern North America. *Bulletin of the Torrey Botanical Club* 71:166–74. In spite of its date, this is a very nice summary; well written, good descriptions.

Butterworts
(*Pinguicula*)

Casper, S. J. 1962. On *Pinguicula macroceras* Link in North America. *Rhodora* 64:212–21. Good summation of this possible new species split from the *P. vulgaris* taxon.

Godfrey, R. K., and Stripling, H. L. 1961. A synopsis of *Pinguicula* in the southeastern United States. *The American Midland Naturalist* 66:395–409. Very good summary paper with superb, complete descriptions, keys, excellent line drawings, and references.

Bladderworts
(*Utricularia*)

There is a paucity of good general articles on this genus in North America, but we understand that this will be corrected in the future. Of the references in the general section, Lloyd has an excellent anatomical discussion with many fine line drawings. The photos of *Utricularia* in the Rickett books are accurately labeled.

Ceska, A., and Bell, M. A. M. 1973. *Utricularia* in the Pacific northwest. *Madrono* 22:74–84. A particularly useful review, since five species are completely described, and all of these occur in the east as well.

Kondo, K. 1972. A comparison of variability in *Utricularia cornuta* and *Utricularia juncea*. *American Journal of Botany* 59:23–37. A very thorough comparative description of these two species.

Kondo, K. (with additional commentary by Peter Taylor). 1973. A key for the North American species of *Utricularia*. *Carnivorous Plant Newsletter* 2:66–69. An excellent, updated, easy-to-use key with difficult points illustrated.

Reinert, G. W., and Godfrey, R. K. 1962. Reappraisal of *Utricularia inflata* and *U. radiata*. *American Journal of Botany* 49:213–20. Another good comparison of two similar species.

Carnivorous Plant Newsletter

Carnivorous Plant Newsletter is a recently conceived quarterly publication for those who have a serious interest in the subject and is intended for nonprofessional as well as professional botanists. CPN features news, short notes, photos, and reviews of recent literature and has a seed and plant exchange for subscribers. For additional information write one of the co-editors: J. A. Mazrimas, 329 Helen Way, Livermore, Calif. 94550, or D. E. Schnell, Rt. 4, Box 275B, Statesville, N.C. 28677.

Glossary

Actinomorphic. Radially symmetrical.

Active trap. A carnivorous plant trap in which a movement of plant parts takes place during the trapping process.

Ala. Literally, "wing"; a broad, bladelike expansion of the axial margin of a pitcher leaf.

Amphiploidy. A process by which a new species develops from a hybrid plant. The chromosome number of the hybrid doubles and the plant is capable of maintaining its characteristics during sexual reproduction with like plants.

Anther. The tip portion of a stamen, which produces pollen.

Anthesis. The period in which a flower expands and/or pollination can take place

Asexual reproduction. A form of reproduction involving only one parent plant and thus no exchange of genetic material; e.g., budding, cuttings, bulb divisions, etc.

Backcrossing. A reproductive cross between a hybrid and one of its parent plants.

Beard. A confluence of plant hairs on the palate of a flower.

Binomial nomenclature. The modern system of biological classification whereby each living organism bears a two-word name corresponding to its genus and species.

Bog. A freshwater, constantly moist or wet area dominated by mosses and herbaceous plants.

Bract. A small modified leaf structure, which, in flowers, is located below the calyx.

Bracteole. A small bract.

Calyx. A collective term for all the sepals of a flower.

Chasmogamous. A term applied to flowers that open or expand fully during anthesis.

Cleistogamous. A term applied to flowers that open only partially during anthesis.

Clone. In botany, a group of plants that all bear the same genetic composition, having been borne of one plant by repeated asexual reproduction.

Closing trap. A carnivorous plant trap in which two identical trap halves approximate and thus incarcerate the plant's prey.

Column. In the context of pitcher plant leaves, the structure supporting a lid or hood.

Corolla. A collective term for all the petals of a flower.

Cross-pollination. The exchange of pollen in sexual reproduction between two different flowering plants.

Cuneate. Wedge-shaped (a term applied to leaves).

Cuticle. A water-impermeable, waxy coating of some plant surfaces.

Door. In *Utricularia*, the veil of tissue that closes a trap opening.

Ensiform. Sword-shaped (a term applied to leaves).

Enzyme. A chemical substance that speeds a chemical reaction without itself changing or becoming a component of the reaction.

Family. A closely related group of genera. A family may have only one genus, but classification is at the same level as other families with two or more genera.

Fenestrations. Depigmented, windowlike areas of plant tissue, also known as *areolae*.

Filiform. Threadlike (a term applied to leaves).

Fimbriate. Feathery, or very finely divided.

Flypaper trap. A carnivorous plant trap in which the

victim is ensnared by sticky, mucilagenous secretions.

Fusiform. Thickened in the middle but tapering smoothly toward each end.

Gemmae. Buds formed by vegetative reproduction in a small cuplike structure from which they are shed.

Genus. The first word or more inclusive portion of a binomial name (pl. *genera*).

Grass-sedge bog. A sandy bog dominated by grasses and sedges with scattered longleaf pines.

Hibernaculum. A winter bud from which plants will arise with the return of proper growing conditions (pl. *hibernacula*).

Hood. A pitcher leaf appendage that usually (or derivatively) hangs over the pitcher opening. Also called a *lid*.

Hybrid. Generally, a plant resulting from a cross between two species.

Keel. A ridge on a pitcher plant trap shaped roughly like the keel of a boat.

Lid. *See* Hood.

Marl bog. A bog in which the "soil" is alkaline marl with calcium carbonate.

Marsh. A tract of wet land, usually with fresh, salt, or brackish water to some depth, dominated by taller grasses and reeds.

Morphology. In botany, the form and nonmicroscopic anatomy of plants.

Obovate. Somewhat oval; a term applied to a leaf or petal which is attached at the narrow end so that the distal end appears broader.

Ovary. The lowermost portion of the pistil, in which eggs develop; the ovary will become the seed capsule after fertilization.

Palate. A prominence on the lower lip of a sympetalous, usually zygmorphic flower.

Passive trap. A carnivorous plant trap in which no plant movement occurs as an integral part of the trapping process.

Pedicel. A stalk supporting only a single flower.

Peduncle. The supporting stalk of one or several flowers.

Petal. The often colorful, form-giving, leaflike portions of a flower located above the calyx.

Petiole. A leaf stalk.

Photosynthesis. The synthesis of sugars from carbon dioxide and water by green plants with the participation of chlorophyl.

Phyllodia. Leafblade-like structures that are probably expanded or widened petioles.

Pistil. The female reproductive portion of a flower, in which seed will form.

Pitfall trap. A carnivorous plant trap into which the prey falls and cannot exit.

Polymorphism. The condition in which plants of the same species (or subspecific classification) have much variation in form.

Primordia. Primitive or undeveloped structures such as those antecedent to leaves.

Quadrifid. Having four parts or branches.

Raceme. A type of inflorescence in which there is a central stalk with the flowers attached by pedicels.

Reticulate. Netlike.

Rhizome. An elongate underground stem, which runs approximately parallel to the surface of the ground, from which branchings may arise.

Saccate. Saclike.

Savannah. In the sense used in this book, a sandy bog with short grasses and sedges and widely scattered longleaf pines.

Scale. A thin, membranous, colorless, often brittle degenerate leaflike structure.

Scape. A long, naked (without bracts, bracteoles, etc.) flowering stem arising from the ground, usually supporting one flower or a tight cluster of flowers at the very top.

Self-pollination. The pollination of a stigma with pollen from the same flower.

Sepal. A flower part situated just below the petals. The sepal is usually green, but if the flower is technically without petals, the sepal may assume the form and color of a petal.

Sessile. Set immediately upon another structure without an intervening stalk, as a *sessile* leaf or gland.

Sexual reproduction. A form of reproduction in which some exchange of genetic material occurs between two organisms.

Species. The second word or most specific part of a binomial name (pl. *species*).

Sphagnum bog. A bog dominated by *Sphagnum* mosses.

Spur. In floral morphology, an elongate, closed appendage of the corolla of a sympetalous flower.

Stamen. The male reproductive structure of a flower, consisting of the anther and its supportive structure, the filament.

Stigma. The sticky, pollen-receptive, often knobby top portion of the pistil.

Stolon. A runner, or any basal branch that forms roots and gives rise to an independent plant.

Stratification. In horticulture, the process whereby seeds are exposed to a period of damp cold before they will germinate.

Style. The often columnar structure of the pistil between the stigma and ovary.

Swamp. A freshwater area, with water to some depth, dominated by trees.

Sympetalous. Having fused or joined petals.

Syngameon. A specialized evolutionary term referring to plant populations intermediate between the species level and extreme variants of the same species.

Threshold. In *Utricularia*, the thickened surface against which the edge of the door rests.

Trapdoor trap. A carnivorous plant trap in which an appendage closes over an opening and incarcerates the plant's prey.

Turion. In this book, a hibernaculum, but used mainly in reference to *Utricularia*.

Vegetative apomixis. A form of asexual reproduction in which plantlets bud from flower parts, including sepals, petals, stamens, and pistil.

Vegetative reproduction. *See* Asexual reproduction.

Velum. In *Utricularia*, a membranous structure for secondary closure; it rests below the door on the threshold.

Zygomorphic. Bilaterally symmetrical.

Derivations of Scientific Names

Understanding the meanings of latinized binomial names used in biology is an aid in learning to use them and in remembering them. We have divided the latinized carnivorous plant names used in this book into two lists below: a list of generic names and a second list of species, subspecies, and forms. In each case we have attempted to provide the best definition, distilled from several botanical dictionaries and our own extensive reading. It is easy enough to provide a "translation" based on Latin or Greek roots, but a biologist is not compelled to explain why he named a plant or animal as he did, and indeed few have. Many original descriptions are lost in antiquity. Consequently, a little detective work and imagination are required to see the application of a particular name to a particular plant. In some cases, we have not the faintest idea why a certain name was chosen, even though we may translate it, and we will say so in those instances. One other word of caution: Latin or Greek language scholars may take exception to some of the botanical meanings below. All we can say is that, in those instances, latinization has been subjugated to botanization!

Generic Names

Darlingtonia. Named after Darlington; in this case, Dr. William Darlington, a nineteenth century naturalist from Pennsylvania.

Dionaea. This name must have the most romantic origin of all. Venus' mother was Dione, and of course Venus was goddess of love who enthralled and beguiled all men. The botanist who came up with this generic name must truly have been beguiled by this little plant.

Drosera. From the Greek, meaning dewy, referring to the secretions at the tips of glandular hairs.

Pinguicula. Literally, "little fat one," referring to the greasy appearance and texture of the leaves of this genus.

Sarracenia. Many early botanists were physicians, as was Dr. Michel Sarrazin (1659–1734), who lived in Quebec and sent specimens of what is now known as *Sarracenia purpurea* to France. His name is now immortalized in this genus.

Utricularia. Literally, little bag or sac, referring to the bladderlike traps of this genus.

Species, Subspecies and Forms

alata. Winged, referring to the prominent ala of *Sarracenia alata*.

amethystina. Reddish-purple in color, like the gem amethyst, referring to the flower color of this *Utricularia*.

anglica. Of England, where *Drosera anglica* occurs commonly and was originally described.

biflora. Two-flowered. The scape of this *Utricularia* commonly, but not always, bears two flowers.

brevifolia. Short-leafed, an apt description for this *Drosera*.

caerulea. Dark blue, here referring to the flower color of this *Pinguicula*.

californica. Of California.

capillaris. Literally, hairlike. The etymology is unclear, but this possibly refers to the many glandular hairs on the leaves of *Drosera capillaris*.

cornuta. From the Latin, meaning horned, referring to the prominent spur of *Utricularia cornuta* (cf. *mac-*

roceras and *microceras* below, which are of Greek origin).

fibrosa. Having prominent fibers, likely referring to the fibrous mats of stems of *Utricularia fibrosa*.

filiformis. Threadlike, an apt description of the leaves of this *Drosera*.

fimbriata. Fringed, describing the margins of bracts and sepals of *Utricularia fimbriata*.

flava. Yellow, which could describe either the flower petals or pitcher tops of some forms of *Sarracenia flava* (cf. *lutea* below, also derived from the Latin and meaning yellow).

floridana. Of Florida, where *Utricularia floridana* most commonly, but not exclusively, occurs.

foliosa. Very leafy or full of leaves, probably referring to the many photosynthetic branchings of *Utricularia foliosa*.

geminiscapa. Literally, having twin scapes. I have no idea how this was applied to *Utricularia geminiscapa* unless someone mistook the rather arching pedicels for scapes.

gibba. Swollen. The application to this *Utricularia* is unclear, unless it refers to the palate, which is no more swollen than that of any other member of the *U. fibrosa* group.

heterophylla. Having a leaf variation, here referring to the pigment variation (the lack of red) of this northern *Sarracenia purpurea*.

inflata. Inflated or swollen, referring to the air-filled arms of the flotation structure of this *Ultricularia*.

intermedia. Obviously intermediate, although we do not know with respect to what in this *Drosera* and *Utricularia*. The reference may be to leaf length in the former and flower size in the latter.

ionantha. Violet-flowered, referring to the color of the flower of this *Pinguicula* or its superficial resemblance to the flower of a violet.

jonesii. Commemorative of Jones, in this case Frank Morton Jones, an entomologist who studied insect associates of pitcher plants.

juncea. Rushlike, probably referring to the appearance of groups of preflowering scapes in wet sand overlaid with shallow water where this *Utricularia* grows.

leucantha. White-flowered, one of the characteristics some would use to separate the populations of this species of *Drosera* from those of *D. brevifolia*.

leucophylla. White-leafed, appropriate for this white-topped *Sarracenia*.

linearis. Linear, as is the elongate leaf with parallel sides in this *Drosera*.

lutea. Yellow (cf. *flava* above), referring to the flower color in this *Pinguicula*.

macroceras. Literally, having a large or long horn (*-keros*, Gr., for horn; cf. Lat. *cornuta* above), referring to the longer spur supposedly characteristic of these populations of *Pinguicula* that some would set aside from *P. vulgaris*.

macrorhiza. Having a long root, but this and other species of *Utricularia* are rootless, so the reference must be one incorrectly applied at an earlier time to the very long stem of this plant.

microceras. Having a short or small horn (cf. *macroceras* above), referring to the unusually short-spurred far northern variant of this western species of *Pinguicula*.

minor. Smaller; perhaps appropriate for the *Utricularia* so named, but not for some of the populations of *Sarracenia* we have seen.

muscipula. Fly-catching; eminently appropriate for the Venus' flytrap.

ochroleuca. Yellowish-white. The flower of this yellow *Utricularia* is somewhat paler than most.

olivacea. Olivelike. This application is lost in history. It may refer to the prominent double ovary (rare in

Utricularia) or possibly even to the swelling, maturing green seed capsule.

oreophila. Mountain-loving, referring to the habitat of this *Sarracenia* in northeastern Alabama.

planifolia. Flat-leaved. While curled on the very edges, the large leaves of this *Pinguicula*, when growing in full sun, are flatter than those of most other species.

primuliflora. Primrose-flowered, referring to the superficial resemblance the flower of this *Pinguicula* has to primroses.

psittacina. Parrotlike. A sideview of the pitcher of this *Sarracenia* discloses a good case for this epithet.

pumila. Dwarf; this the smallest southern *Pinguicula*.

purpurea. Purple, referring to the deep maroon veins or flower parts of *Sarracenia purpurea*, and to the pale purple flower color of the *Utricularia* species.

radiata. Rayed or radiate, the arrangement of the spokes of the flotation structure in this *Utricularia*.

resupinata. Upside down, or apparently so, as the general appearance of the flower of this *Utricularia*.

rotundifolia. Round-leafed. Although generally the blade is more obovate than round in this *Drosera*, the epithet is appropriate.

rubra. Red. Applicable to pitcher or flower color in this *Sarracenia*.

simulans. Similar to or resembling; a relative term possibly referring to this *Utricularia*'s resemblance to *U. fimbriata*.

standleyae. Commemorative of Standley; in this case, Paul E. Standley, a noted systematic tropical botanist of the early part of this century.

subulata. Awl-shaped. The exact reference is not known but is likely related to the shape of the flower spur.

typica. Typical.

venosa. Veined.

villosa. Softly hairy; referring to the pubescence on the lower portion of the scape of *Pinguicula villosa*.

vulgaris. Common.

Some Prefixes

bi-. Two.

brevi-. Short or abbreviated.

gemini-. Twin.

hetero-. Other.

leuc-, leuco-. White.

macro-. Large; sometimes long.

micro-. Small.

ochro-. Yellow (ochre).

oreo-. Mountain.

plani-. Flat.

primuli-. Primrose (genus *Primula*).

rotundi-. Round.

Some Suffixes

-antha. Flower.

-ceras. Horn.

-flora. Flower.

-folia. Leaf.

-leuca. White.

-phila. Loving, or affiliated with.

-phylla. Leaf.

-rhiza. Root.

-scapa. Scape.

Index

This book is set in 10 on 13 Palatino
Book design by Virginia Ingram
Photographs by the author
Drawings and maps by Bruce Tucker
Composition by Heritage Printers, Inc.,
Charlotte, North Carolina
Printing by Lebanon Valley Offset Company, Inc.
Annville, Pennsylvania
on Warren's Lustro Offset Enamel Dull, White, 80–lb.
Binding by Optic Bindery, Baltimore, Maryland
The cover is Holliston's Roxite B Linen Finish
Endpapers are from Process Materials Corp.